Merton, Henry A.
Your gold and silver

YOUR GOLD AND SILVER

AN EASY GUIDE TO APPRAISING HOUSEHOLD OBJECTS, COINS, HEIRLOOMS, AND JEWELRY

HENRY A. MERTON

Your Gold and Silver

Your Gold and Silver

An Easy Guide to Appraising Household Objects, Coins, Heirlooms, and Jewelry

HENRY A. MERTON

COLLIER BOOKS
A Division of Macmillan Publishing Co., Inc.
New York

COLLIER MACMILLAN PUBLISHERS
London

Macmillan Publishing Co., Inc.
866 Third Avenue, New York, N.Y. 10022
Collier Macmillan Canada, Ltd.

Library of Congress Cataloging in Publication Data
Merton, Henry A.
Your Gold and silver.
Includes index.
1. Gold. 2. Silver. 3. Goldwork—Valuation.
4. Silverwork—Valuation. 5. Silver coins—Valuation. 6. Jewelry—Valuation. I. Title.
HG261.M47 332.63 81-4966
ISBN 0-02-077410-9 AACR2

10 9 8 7 6 5 4 3 2 1

Printed in the United States of America

dedicated to
Sol
and all the Pinsleys

Contents

Illustrations and Charts

Preface

The average shopper, when purchasing most items, exercises great care in determining how he or she spends money.

We lift up the hood of a used car, kick the tires, slam the door; or spend many minutes at the canned goods section of the market comparing different brands of the same food.

But when it comes to purchasing quality jewelry, fine watches, silver sets, and other items in the precious-metals field, most of us depend on the integrity of the manufacturer, jeweler, or merchant selling the goods.

This is not to say that we don't compare one ring against another, or check many sets of silverware, but in doing so we usually satisfy our desire for an attractive design, or possibly are shopping for the best price. Most of us do not compare values in relation to the actual value of the precious metal content.

Unlike two different brands of the same size can of green beans, two equally sized rings might be a world apart in actual precious metal value.

This lapse in the normal buying habit is not due to the individual's laxity, but to a lack of knowledge. Few persons have more than a cursory acquaintance with the names, precious metal content designations, weight system, and many of the other (sometimes) confusing terms applied to jewelry, sterling silver pieces, coins, and the like.

Because of this general lack of knowledge, combined with a genuine need for it now that people are buying and selling gold and silver, I wrote this book.

With a little time, the fundamentals of the field of precious metals can easily be understood. Words such as solid gold, karat gold, vermeil, 14KGF, coin silver, bullion, gold flash, and all the others, will lose their mystery. And with this veil of mystery lifted, you can approach your dealings in the precious-metals field with the same confidence you have when buying your weekly groceries.

EXISTING INFORMATION ON GOLD AND SILVER

The need for this book was further brought home to me when I tried to find the answers to some timely questions on gold and silver. I visited several libraries but found no wide-ranging book that covered the basic needs of anyone who wants to buy or sell gold or silver. Even an encyclopedia was of little help. There were good books on specific subjects, such as numismatics and investments, but a good, general publication was not to be found. With the ever-increasing interest in precious metals, an informative work was clearly called for.

Further, available information was not always consistent. For example, in several books a troy ounce was variously defined as equaling 31.104 grams, 31.1033 grams, and 31.1035 grams. Now, that may not seem like a major difference, and it actually isn't. (With gold at $500 per troy ounce, the discrepancy between an ounce equaling 31.104 grams and an ounce equal to 31.1033 grams amounts to 35¢ in one ounce.) The actual difference in value was not the important point; what is important is the fact that different gram values were given by authoritative sources. The *International Dictionary of Physics and Electronics* showed that one troy pound equals 373.25 grams. Using an eight-place pocket calculator, 373.25 ÷ 12 troy ounces = 31.104166. Based on this, all calculations in this book use the equivalent, 1 troy ounce equals 31.104 grams.

I also found much misleading information in current publications. Several articles and books about gold jewelry stated that, for instance, a ring stamped 14K has a fixed gold content. That is not necessarily true. A ring plainly marked 14K may in fact be only 13½K, or 13K. It all depended on the policy of the specific manufacturer of the ring.

Many, if not most, dealers (or *gatherers,* as they are called by many in the trade) purchasing scrap silver and gold are not aware of the laxity in the old stamping laws that allowed a manufacturer to put up to one full karat *less* than was stamped on the manufactured article. Of course, at the low prices (in comparison to the spot precious-metals prices) the gatherers have paid, and are most often still paying to people selling their valuables, a difference of one karat will be of little notice. The gatherer's profit margin is well protected.

The information contained in this book will be of use to persons buying or selling precious metals; to those interested in learning the meanings of the terminology used by the silver and gold industries; to potential buyers of bullion or numismatic coins; and to owners of jewelry or household objects who can be more appreciative of their possessions by understanding their intrinsic value as well as artistic worth.

A lack of knowledge of precious metals has created a problem for the public. In the late 1970s and into the 1980s, people have been selling, and will continue to sell, their unwanted (or overly valuable) jewelry, silverware, coins, sterling silver, and many other forms of personally owned precious metals. These sellers are drawn by offers of high profit. Unfortunately, most people approach the precious-metals-buying operation with blind trust. This is not meant to suggest that gatherers are dishonest, but that you should exhibit the same caution you use in any other kind of transaction.

The most common precious metal you are most apt to be bamboozled over is sterling silver. Don't let the prospect of making some ready cash by selling the tea set or candelabra you haven't used for years blind you to their true worth. Quite often you could make a better deal by taking your items to a jeweler, an antique dealer, or a person who deals in collectibles. Of course, many of these people also have become buyers of silver and gold. There are many discriminating and conscientious dealers, however, who disdain the fast buck. Shop around. Don't be fooled by the splash of newspaper ads, sandwich-board signs, or window advertisements. When gold was fluctuating from $600 to $800 a troy ounce, the average highest price paid for scrap gold by the gatherers was no higher than $400. Many other buyers were paying no more than $200 an ounce! Many gatherers made, and are still making, enormous profits from people who have no knowledge of the true value of their precious metals. Many gatherers bought sterling-silver pieces, silverware, jewelry, tea sets, trophies, wall plaques, paying their bulk-silver prices, then they polished the items and sold them for their true artistic worth, many times the raw-silver value. The seller is mesmerized by the prospect of getting in on the big push. Why else would they go to a motel room and deal with a complete stranger, when there are established dealers in that same town who offer just as good, if not better, deals.

The exorbitant profits made by many gatherers should dwindle and the prices offered to the seller should rise, if the seller is knowledgeable about the actual value of precious metals, and if the seller compares this metal value with the artistic, antique, or collectible value as well.

With the information found in this book, you can comprehend how the precious-metal market works. If you believe the best offer you find is still too low, don't sell. If enough of the public does this, the market will become more of a seller's field and the prices paid by the gatherers will improve.

Acknowledgments

The publication of this book is a testimonial to the splendid cooperation and assistance of many individuals, companies, corporations, organizations, associations, and institutions. Appreciation must be expressed to the following for making this work possible: Reita Wojtowecz; Herbert Potoker; Jack Petker; Seymour Saslow; David Cavenaugh; Joseph Canterino; Jerry Roberts; Harold Ostrov; Edward Pilkington; Reita Finnegan; Walter Stroup; David Sexton; Barry Pinsley, certified public accountant; Westbury Alloys Corp; Handy and Harman; Engelhard Industries; E. I. Dupont DeNemours & Co.; Eastman Kodak; Gamzon Bros., Inc.; Republic National Bank of New York; Brinkmann Instruments, Inc.; Ohaus Scale Corp.; Deak-Perera; Manfra, Tordella & Brookes, Inc.; Consolidated Gold Fields Ltd.; Gorham Textron; The Gold Institute; The Silver Institute; The Gold Filled Manufacturers Association; The Jewelers Circular Keystone Magazine; the Saratoga Springs Public Library; the Lucy Scribner Library, Skidmore College; Charles Swick; Margaret Roohan; and any other person or firm I might have inadvertently failed to mention.

I especially want to thank Charles Levine for all his help in turning my manuscript into a finished book.

The person I would most like to thank, however, is no longer jaunting down Lincoln Avenue in Saratoga Springs, New York. Frank Sullivan gave me encouragement and advice at times when I felt less like a writer than a consumer of vast amounts of typing paper. He told me, "My usual comparison between a writer and an author is that an author is a writer who successfully badgers publishers." Frank never had to "badger," at least not after they recognized his genius for humor. Here's to you, Frank.

Your Gold and Silver

Part One

Fundamentals of Precious-Metals Trading

1

Beware of Frauds

The foremost rule in *any* business transaction is: Know with whom you are dealing.

There are reputable dealers in every area of the precious-metals market: jewelers, coin merchants, investment brokers, bankers, and refiners. If you do not know how to contact someone in the area of your interest, check with your banker, the chamber of commerce, or go to the library or a bookstore to find a publication on the subject. See if there is a magazine covering the field; look for an association or organization related to your interest; or contact one of the firms mentioned in this book. In this way you can sensibly get a start in the precious-metals field.

Never react on-the-spot to a telephone or door-to-door solicitor. High-pressure tactics and super salesmanship bilk the public of millions of dollars every year. This is not to say that all such salespeople are dishonest. Make it a rule not to succumb to a stranger selling anything. Even if the deal sounds too good to pass up, tell the caller to come or call back at a later time that you specify. Then take the time to check out the offer. If the salesperson tells you that you have to jump on the bandwagon at that moment or lose out, fine; it wasn't anything you planned for, so it would be no great loss to you.

"Fool's Gold investment-firm records seized." "$100 Million Fraud." The papers are turning up each day more such stories of the public being taken in by fraudulent gold and silver schemes.

Don't be one of the sorry ones.

Because of widespread abuses by gatherers, many local municipalities and some state governments have started to enact laws or ordinances to regulate dealings in precious metals. But legislators have been discouraged from passing strict laws by the rampant abuse of dealers and by an increase in the theft of jewelry and precious items. With the increase in the value of gold and silver, and the ease of selling stolen items, thieves and unscrupulous dealers have prospered. Good statewide, all-encompassing laws on precious-metals dealings are needed now, not in a few years. Citizens can be protected if legislators eschew partisan politics and act now, with the interests of their constituents in mind.

2

Properties of Gold and Silver

GOLD

The chemical symbol for gold is Au.

Gold's atomic number is 79 and its atomic weight is 196.967.

Gold has a relative hardness of 2.5 (on a scale of 10).

Gold melts at 1064.43° Centigrade.*

The specific gravity of gold is 19.3. This means that gold weighs 19.3 times more than an equal volume of water.

SILVER

The chemical symbol for silver is Ag.

Silver's atomic number is 47 and its atomic weight is 107.868.

Silver has a relative hardness of 2.7 (on a scale of 10).

Silver melts at 961.93° Centigrade.*

The specific gravity of silver is 10.50. This means that silver weighs 10.50 times more than an equal volume of water.

3

Abbreviations for Weights and Measures

The abbreviations for weights and measures commonly used in this book are: pound (lb.), ounce (oz.), pennyweight (dwt.), metric gram (g.), grain (gr.), millimeter (mm.).

Don't confuse grain (gr.) with metric gram (g.) Both units are used often in this book.

* *Handbook of Chemistry & Physics* (Boca Raton, Fla: CRC Press, 1979). According to this source, these melting points of gold and silver have been recently recalculated.

The next section discusses the system of weight used in measuring gold and silver.

4

Weights and Measures for Gold and Silver

Gold and silver are weighed according to the troy system of weights. This is important to keep in mind: The prices quoted in newspapers, from worldwide markets, or in advertisements use troy weights. To better understand these weights, let's first compare them with avoirdupois weights, the system used for food, candy, iron, cotton, plastic, shipping, and virtually everything the public commonly uses.

There is only one point at which avoirdupois and troy weights are the same (equal), and that is at their smallest unit, the grain. An avoirdupois grain and a troy grain are equally balanced on a scale. From that point on, the avoirdupois and troy systems differ greatly. *A troy ounce, for example, which is the ounce used to quote the value of gold and silver, is nearly ten percent heavier than an avoirdupois ounce.* An avoirdupois pound, on the contrary, is about twenty-two percent heavier than a troy pound. This can be seen on the chart below by comparing ounces and pounds in the two systems as they convert into grains, the smallest unit common to both systems.

Comparison of the Troy and Avoirdupois Systems

TROY	AVOIRDUPOIS
grain = smallest unit, identical in both troy and avoirdupois	
1 pennyweight = 24 grains	1 dram = 27.34375 grains
1 ounce = 20 pennyweights = 480 grains	1 ounce = 16 drams = 437.5 grains
1 pound = 12 ounces = 240 pennyweights = 5,760 grains	1 pound = 16 ounces = 256 drams = 7,000 grains

Note: Throughout this book, ounce (oz.) refers only to a troy ounce.

Because metric weights and measures are now also commonly used for precious metals, the charts below show how to convert grams to troy ounces and millimeters to inches.

Conversion of Troy Weights into Metric Weights

TROY	METRIC
1 grain	= 64.8 milligrams = 0.0648 grams
1 pennyweight = 24 grains	= 1555.2 milligrams = 1.5552 grams
1 ounce = 20 pennyweights	= 31.1042 grams
1 pound = 12 ounces	= 373.248 grams

Example: To convert grams into troy ounces, divide the gram figure by 31.1042. If you have 572 grams, this equals 572 ÷ 31.1042 = 18.3898 troy ounces.

Common Gram Weights Converted to Troy Ounces

5 grams =	0.1607 ounce
10 grams =	0.3215 ounce
20 grams =	0.6430 ounce
50 grams =	1.6075 ounces
100 grams =	3.2150 ounces
1,000 grams = (1 kilogram)	32.150 ounces

Conversion of Millimeters to Inches

One inch equals 25.4 millimeters, or equivalently, 1 millimeter equals 0.03937 inches (which makes 39.37 inches = 1 meter). To convert millimeters to inches, multiply by 0.03937.

Example: A Krugerrand has a diameter of 32.3 mm., which is 32.3 × 0.03937 inches = 1.2717 inches.

Commonly Used Thickness Designations with Equivalents

Microinch	= 1/1,000,000 inch
Mil	= 1/1,000 inch
Micron	= 1/1,000,000 meter
Millimeter	= 1/1,000 meter
Centimeter	= 1/100 meter

5

Glossary of Words and Terms for Precious Metals

It is unusual to find a glossary at the beginning of a book, but the subject matter of this publication requires a basic knowledge of these words to understand the full meaning of the information contained in these chapters. Special care has been taken to try to make all the information easily understandable. Still, fine gold is different from gold alloy, and you must know that difference.

Acid Test. The act of subjecting precious metals to specific acids, or combinations of acids, to determine the fineness of the precious metal.

Alloy. Metal made by the fusion (combination or mixing) of two or more metals. Any silver or gold of less than 999 purity is an alloy.

Actual Gold (or Silver) Content. The amount of precious metal present in an alloy, given as a percentage, a fineness, or a troy weight.

Assay. Determination of the purity of a precious metal and its base-metal content; the work being done by an assayer.

Avoirdupois. The weight system most commonly used in the United States for cloth, food, steel, paper, and most everything the American public deals with, excepting precious metals and gems.

"*Bag.*" Usually a $1000 bag of coins.

Base Metal. The nonprecious metal that serves as a base for gold-filled, gold-plated, silver-plated, or any precious-metal-covered nonprecious metal.

Base Price. The silver-metal value of one dollar in coin, based on the face value and weight of the coin. Because the face value and weight don't change, the base price never changes.

Bulk Metal. The term used when referring to accumulations of coins, sterling silver, scrap jewelry, etc.

Bullion. Precious metal in negotiable or tradeable shape, such as a wafer, a bar, ingot, or (sometimes) as coins and jewelry.

"Burned." Slang used by clandestine smelters for melting coins.

Carat. A unit of weight for gemstones. See also **karat.**

Cartwheel. Slang for a silver dollar.

Circulation. Coins minted as legal tender and put into public use.

Cash Price. (Same as **spot price**) The price set or fixed by the precious-metals market at the current moment, usually daily, for the immediate settlement of transactions. This price can differ at the same time in different markets.

Coin Gold. The alloy used to make gold coins, which may differ from country to country, or in different coins minted in the same country.

Coin Silver. The alloy used to make silver coins. When used as a designation on silver items in the United States, it must be 900 fine (with the deviations allowed by the United States Stamping Act).

Ductile. The capacity of a metal to be hammered into a thin sheet or drawn into a thin wire.

Daily Gold (or Silver) Price. Same as **cash price** or **spot price.**

Electroplating. The act of using an electrical current to plate or flow and fuse a thin layer of precious metal onto a base metal.

Face Value. The value of a coin, paper money, or other currency as imprinted, stamped, or marked on that unit. Further, the value of an accumulation of money (the face value of ten dimes is one dollar).

Fine (also Fineness). The designation of the purity of a precious metal in relation to 1000 parts. For example, 900 fine gold has 900 parts of pure (fine) gold and 100 parts of an alloying metal or metals, which means it contains 90 percent pure gold.

Fine Gold. Pure gold without any alloy metal to contaminate it.

Fine Ounce. An ounce of pure (actually 999 pure) precious metal.

Fine Weight. The actual weight of the pure gold or silver in a coin, ingot, bar, or other item with a precious-metal content (as opposed to the item's total weight, which includes the weight of the alloying metal).

Fix. To set the price of gold or silver daily (on trading days). The London Gold Market fixes the prices twice daily.

Full Value. The value of a precious-metal alloy before taking into consideration wear, and assaying and refining costs.

Futures. A price set in precious-metals trading for a later delivery, as opposed to the price for immediate "spot" deals.

Gatherer. A person or a firm involved in the purchase of scrap precious metals, who could be a jeweler, a coin dealer, or an antique dealer. Gatherers can be organized for the sole purpose of buying scrap precious metals.

Gold Electroplate. An item on which a thin coating of karat gold has been applied to a base metal by electrical current. The karat-gold finish must be at least seven-millionth inch thick.

Gold-Filled. Refers to an item on which a layer of karat gold has been applied to a base metal using heat and pressure. The gold-filled item is then drawn or rolled to a specific thickness. The karat-gold content must be at least one-twentieth of the total weight of the item.

Gold-Flashed (or Gold-Washed). A thin film of gold applied to a base metal, as in electroplating, but with less than seven-millionth inch thickness of karat gold.

Gold-Plate. (Also **gold-overlay** or **rolled-gold-plate**). Basically the same as **gold-filled,** but the gold content can be less than one-twentieth of the total weight of the item if so marked.

Gold-Overlay. Same as **gold-plate**.

Gold Standard. The term used to designate the monetary standard of a country when all the paper money it issues is backed by a gold reserve.

Grain. The smallest unit of measure in both the troy and avoirdupois weight systems, equivalent to 64.8 milligrams.

Gram. A metric weight equal to one-thousandth of a kilogram and commonly used in precious-metals dealings.

Hallmark. The marks put on bullion by the refiner to show weight, fineness, manufacturer, and serial number.

Heavy Gold Electroplate. Similar to **gold electroplate,** but the gold finish must be at least one-ten-thousandth inch thick.

Intrinsic. Belonging to an item by its very nature: The *intrinsic value* of a coin is the worth of its metal content rather than its face value or numismatic value.

Issue. The coins or currency turned out or released at a specific time. The date quite often defines the issue, but not always. Sometimes one date is issued at several different times, or coins are restruck with the original dates.

Karat. Measurement of purity used in showing the fineness of gold. One karat is one-twenty-fourth pure gold. Thus, 24-karat gold is pure (999 fine).

Karat Gold. Gold usually used in jewelry manufacturing that is at least ten karats or better. United States law requires metal to be at least 10K minimum or the manufacturer cannot call it "gold." However, by a quirk in the United States laws, this requirement is subverted (until October 1, 1981, when the law will change) by another law that allows karat gold under certain circumstances to be up to one karat less than marked.

Layered Metal. As used in this book, a metal item in which gold or silver is used as a finishing surface on a base metal, such as gold or silver plate, as opposed to an alloy where the metals are actually mixed or combined together.

Liquid Gold. A liquid created by the mixture of gold with certain chemicals. Primarily used as a surface coating, such as on glass for reflective qualities.

Lot. A trading designation for bullion coins, comprising twenty coins of one denomination.

Malleable. The characteristic of a metal allowing it to be hammered into a thin sheet without cracking. Gold is the most malleable of all metals.

Mark (Marking). The stamping on silver and gold products that shows fineness, manufacturer's trademark, type, and other indications as required by United States laws. Also called **hallmark.**

Melt and Assay. The policy of a refiner to melt and test an alloy before making a settlement on purchases of precious metals.

Metric. The decimal system of measurement involving grams, kilograms, millimeters, and meters—now widely used with precious metals.

Mint. The plant where coins are manufactured.

Minting. The act of manufacturing coins in a mint.

Numismatics. Relating to the hobby or occupation of collecting coins and related items.

Numismatist. A coin collector.

Operation. As used in this book, the business location of a gatherer.

Option. An agreement to buy or sell a commodity or stock at a later date.

Ounce. A weight designation for both troy and avoirdupois systems. They are not equal, however.

Pennyweight. One of the more common troy-weight designations found on a precious-metals scale.

Plumb Gold. This term applies to karat gold that is "plumb" with the karat mark stamped on the product. Fourteen-karat plumb gold stamped *14K* is 14K. It does not deviate as has been allowed by United States law, down to 13½K or 13K.

Pound. A weight designation for both troy and avoirdupois systems. They are not equal in balance, however.

"Pot." Slang for a smelting furnace.

Premium. The value of a precious-metal item, such as a coin, over and above the intrinsic value of the metal content.

Pure Gold. Gold of 999 fineness or 24-karat gold, with no alloying metal.

Pure Silver. Silver of 999 fineness, with no alloying metal.

Refined. Precious metals that have been melted and worked to separate the precious metals from the alloying metals.

Refinery. Plant where precious metals are refined.

Restrike. When a government takes an out-of-issue coin and produces a new minting using the original dies.

Rolled Gold Plate. See **gold plate.**

Scale. The weighing device used to determine the weight of an item.

Settlement. The act of a buyer paying the seller for a purchase.

Solid Gold. One of the most misleading designations the United States government allows manufacturers of precious-metal products to use. It means simply that the gold product is not "hollow." The gold item could have as little as nine karats (with deviation from 10K) and still be stamped *solid gold.*

Speculate. The act of dealing in precious metals for resale at a later date, rather than dealing for an immediate transaction.

Spot Price. See **cash price.**

Stamping. The mark put on a precious-metal product which is within the jurisdiction of the United States Stamping Act. See **mark (marking).**

Sterling Silver. The name applied to objects of jewelry, housewares, etc., that have a fine-silver content of 92.5 percent (925 fine). The most common alloy dealt with in the daily gatherer's business.

Tael. A Chinese system of weights used for precious metals: 1 tael equals 1.2034 troy oz. of pure gold.

Tolerance. The permissible variation, above and below the actual weight of an object being weighed. A tolerance example is (plus or minus) .05 percent of an item being weighed.

Troy Weight. The system of weight primarily used in the United States for precious metals. The weight used throughout this book.

Vermeil (pronounced by jewelers Vair-May). An ambiguous word incorrectly thought to apply to gold-covered silver (sterling). In fact, however, vermeil can be applied to many metals, including nonprecious ones. Federal regulations for vermeil are under consideration and could soon be passed.

Wafer. The designation for bullion produced in a small, thin form.

6

Calculating the Value of a Gold or Silver Item

The values of the variety of precious metals covered in this book are all calculated in basically the same way. The same principles apply to all gold and silver objects. The tools you need to evaluate intelligently an offer to buy precious metals are the information in this book, a pad and a pencil, a pocket calculator, and a publication quoting the spot silver or gold price of that day. With all of these in hand, you will be able to calculate the value of your gold or silver item and compare it with the price offered by a buyer—or conversely, if you want to buy gold or silver, with the price offered by a seller. In this way, you can better decide whether you are being offered a fair price for these valuable items.

This book covers all types of gold and silver items, from sterling silverware to gold bullion. Various coins are also discussed, including United States silver and gold, other worldwide tender in silver and gold, bullion coins, and other precious-metal coins and items. Once you understand how to calculate their value, you can do so for all gold and silver objects, regardless of the country of origin, weight, or fineness.

In this section, we will discuss in general how to appraise gold or silver. The explanation is followed in the next section by an example taken from an actual advertisement. Each of the offers in the ad is used as the basis for our calculations. They are examples of the process you should use before buying or selling precious metals. The advantage of a newspaper advertisement is that you can determine at home if the prices being offered are fair before going to the buyer or seller.

Let us first consider the silver coins covered in this book, because the charts in the relevant sections give the actual silver content for different denominations of various types of silver coins and calculations are straightforward.

Follow these steps for silver coins:

1. Look up the specific chart, where you will find the actual silver content for given denominations.

2. Multiply the weight of pure silver by the spot metal price. This gives the full market value of the pure or fine silver in the coins.

Example: You have $100 face value worth of Canadian silver coins. The spot silver price for that day is $15 a troy ounce. The weight given on the Canadian silver coin chart (on page 116) shows that $100 of these coins contains 60 ounces of fine silver. Therefore, the market value for your coins is determined by the calculation: $15 per oz. × 60 oz. fine silver = $900. The fine-silver content of your $100 worth of Canadian silver coins has a value of $900. Subtracting $11 for refinery charges and $18 losses (for more details see the section on "Assaying and Refining"):

$$\$900 - \$29 = \$871$$

At a spot silver price of $15 an ounce, the full market value of your $100 worth of Canadian silver coins is $871.

If a dealer offers a payment of five to one, you multiply the face value of the coin by five, which means that the dealer will pay you $500 ($100 × 5) for your coins. Subtracting this amount from the full market value of the coins, $871 − $500 = $371, we have the dealer's profit, which you can see is quite large.

In the event that you have a coin not covered by any of the charts in this book, you must find some reference that gives the fine-silver (or gold) weight. In some cases, you will not find the precious-metal weight, but the full weight of the coin and its fineness. In this case follow these steps:

1. By placing a decimal point in front of the fineness, you have the percentage (in decimal form) of the precious-metal content.

$$\text{FINENESS} + \text{decimal point in front}$$

$$= \frac{\text{FINENESS}}{1000}$$

$$= \text{decimal percentage of precious-metal content}$$

2. Find the total weight of the coin and convert it into troy ounces. If the weight is given in grains, divide by 480 (480 gr. equals 1 troy oz.). If the weight is in grams, divide by 31.104 (31.104 g. equals 1 oz.). This gives you a troy-ounce figure to work with. Multiply the ounce weight and the fineness; this calculation gives the pure-gold or pure-silver content.

$$\text{WEIGHT IN TROY OUNCES} \times \text{FINENESS}$$

$$= \text{GOLD OR SILVER CONTENT (in troy oz.)}$$

3. Multiply the result in number 2 above by the spot metal price and you have the market value of the gold or silver in the object.

$$\text{GOLD OR SILVER CONTENT (in troy oz.)} \times \text{SPOT METAL PRICE}$$

$$= \text{MARKET VALUE}$$

Example: A reference work states that a South African Krugerrand has a total weight of 33.96 grams and a fineness of $916\frac{2}{3}$.

1. The fineness means that the gold content is .91666.
2. The full weight converts to troy ounces as follows:

$$\frac{33.96}{31.104} = 1.0918 \text{ troy oz.}$$

The weight multiplied by the fineness yields the gold content:

$$\begin{array}{r} .91666 \\ \times\ 1.0918 \\ \hline 1.0008 \end{array} \text{ oz. fine-gold content (or 1 full ounce)}$$

3. If the spot metal price of gold is $700, then the coin has a full value of $700.00. Obviously in this case, when a coin's fine content is so close to one ounce, there is no need to multiply. By convention the coin is considered to be intrinsically worth the spot price of $700.

Example: A Swiss 20-Franc gold piece has a fine-gold content of 0.1867 oz. At a spot gold price of $700 an ounce, this coin is worth $700 × 0.1867 = $130.69.

The method of figuring the value of any item is the same, whether it is gold or silver. Because silver is so much lower in value, however, you usually buy or sell bullion silver coins in bags of at least $100 face value, and more often in $1000 bags. Gold generally is sold one coin at a time.

If you find differences between reference works giving exact coin weights or other information, you have to evaluate which source is more accurate. For example, a book published by an expert says that a $1000 bag of American silver coins is smaller than a volleyball and weighs about 55 pounds, with its refined-silver content about 720 oz. In fact, the refined-silver content is 723.37 troy oz., while the total weight is 66.97 troy lb., or 55.11 avoirdupois lb. In one sentence, this expert confused troy and avoirdupois weights. Is it any wonder that the public is confused about precious metals?

If you have a gold object or coin with a karat designation, you must convert this to fineness. As a general rule, when the karat designation is precise, and the manufacturer has not taken advantage of laws that allow a deviation up to one karat, each true karat means one-twenty-fourth part, or 041⅔ fineness of gold.

One karat = $\frac{1}{24}$ part = $41\frac{2}{3}$ fineness

See the chart on page 37 for converting karat into fineness, with deviations of up to one karat. Once you have the fineness, you can then calculate the value of the gold piece as shown earlier in this section.

7

Evaluating Advertisements

The advertisement below was taken from a newspaper. It will be evaluated in the following pages, so you can more fully understand a gatherer's deal. This example is presented here to give you an idea how to analyze an offer. This advertisement ran when gold was selling at $350 an oz. and silver at $42 an oz.!

The section in the center reads: "BUYING GOLD, SCHOOL RINGS, [etc.]." This part of the offer cannot be evaluated because the dealer doesn't give any prices with that category. To determine the fairness of the offer on gold, you would look up the method of calculating found on page 43. This method would be used in conjunction with the spot price for gold. First, you would convert karat designations into fineness with the chart found on page 37.

"U.S. Gold Coins"

(See the chart on page 57.)

"$1— paying $140"

The weight of fine gold in a $1 gold coin is 0.048375 oz. Price offered being paid based on fine-gold content: $2894.05 per oz. ($140 ÷ .048375).

"$2.50—paying $140"

Weight of fine gold is 120833, 0.1209375, or .1289062 oz., depending upon which coin of three. Price based on fine-gold content: $1158.62 per oz. (on 0.120833-oz. coin).

COINS WANTED

Dimes, quarters and half dollars prior to 1965.
Paying $19.00 on the dollar.
Canadian Silver Coins—dimes, quarters and half dollars prior to 1967.
Paying $10.00 on the dollar.
Kennedy Half Dollars 1965 thru 1970.
Paying $2.50 a piece.
Silver Dollars to 1935.
Paying $20.00 a piece (fine or better)
Paying $13.00 an ounce for Sterling Silver.
Jewelry, Knives, Forks, Spoons, Rings, Plates, etc.

JEWELRY	BUYING	BRACELETS
DENTAL	GOLD	NECKLACES
CHAINS	SCHOOL RINGS	RINGS

U.S. GOLD COINS

$1	paying **$140**
$2.50	paying **$140**
$3	paying **$260**
$5	paying **$140**
$10	paying **$200**
$20	paying **$400**

Prices Subject to Change!

"$3— paying $260"
Weight of fine gold is .1478124 oz. Price being paid based on fine-gold content: $1758.98 per oz.

"$5— paying $140"
Weight of fine gold is .2416667, 0.241875, or 0.2578124

oz., depending on which coin of three. Price based on fine-gold content: $579.31 per oz. (for the 0.2416667-oz. coin).

"$10— paying $200"
Weight of fine gold is 0.4833334, 0.48375, or 0.5156249 oz., depending on which coin of three. Price based on fine-gold content: $413.79 per oz. (for the 0.4833334-oz. coin).

"$20— paying $400"
Weight of fine gold is 0.9675 oz. Price based on fine-gold content: $413.43 per oz.

All of the above prices being offered for the gold coins exceed significantly the current spot price of gold. Obviously the dealer is basing his offer on the numismatic or collectible value of the coins, and to more fully evaluate this offer you have to refer to coin-collecting books.

Now we'll evaluate the offer for silver items.

"Silver Dollars to 1935"

"Paying $20.00 a piece (fine or better)"

The dealer is paying 20 to 1. Refer to the chart in the section on the United States silver dollar (page 114). The fine weight of one dollar's worth of silver dollars is 0.773437 oz.; therefore, 0.773437 oz. × $42 (possible spot price) = $32.48.

This dealer is paying $20.00 for $32.48 worth of silver dollars.*

"Paying $13.00 an ounce for sterling silver"

Refer to the section "Sterling Silver" (pages 85–90) and use the method of calculation shown there. This method shows that sterling silver, at 925 fine, has 11.1 oz. fine silver in one troy pound. Since the dealer's price is $13.00 an ounce, you must apply 925 fine to an ounce, that is, 92.5 percent (.925) of one ounce, or .925 × $42 (spot price) = $38.85.

(If the price offered by the dealer is per *pound* of sterling silver, you multiply by 11.1. If the price offered by the dealer is per *ounce* of sterling silver, you multiply by .925.) The dealer is paying $13.00 for $38.85 worth of sterling silver.*

* In the examples above, a further deduction of 2 percent should be made for refinery losses, and about 18¢ per ounce for refining fees (see "Assaying and Refining," below).

"Dimes, quarters, and half dollars prior to 1965"
"Paying $19.00 on the dollar"

The dealer is paying 19 to 1. Refer to the chart in the section on United States silver coins (page 110). The fine weight of one dollar's worth of silver coins is 0.72337 oz. The spot silver price is $42 per ounce; therefore, 0.72337 oz. × $42 = $30.38. This dealer is paying only $19.00 for $30.38 worth of United States silver coins.*

"Canadian silver coins—dimes, quarters, and half dollars prior to 1967"
"Paying $10.00 on the dollar"

The dealer is paying 10 to 1. Refer to the chart in the chapter on Canadian silver coins (page 116). The fine weight of one dollar's worth of Canadian silver coins is 0.6 of an ounce; therefore, 0.6 oz. × $42 = $25.20. This dealer is paying $10.00 for $25.20 worth of Canadian coins.*

"Kennedy half dollars 1965 through 1970"
"Paying $2.50 a piece"

The dealer is paying 5 to 1. In the section on Kennedy half dollars, refer to the chart on page 113. Since $2.50 for a half dollar equals $5.00 for one dollar, the fine weight of one dollar's worth of Kennedy half dollars is 0.29568 oz. Therefore, 0.29568 oz. × $42 = $12.42. This dealer is paying $5.00 for $12.42 worth of Kennedy halves.*

In this case, the dealer, like many dealers, takes advantage of the lack of knowledge relating to the actual silver content of different types of silver coins. Most people are totally unaware of the fineness of the different coins, as well as that of sterling silver. For example, the chart below lists each type of silver with its fine content:

Fineness of Different Silver Alloys

Pure silver	99.9% silver—999 fine
Sterling silver	92.5% silver—925 fine
Coin silver	90.0% silver—900 fine
U.S. silver coin	90.0% silver—900 fine
Canadian silver coin	80.0% silver—800 fine
U.S. silver Kennedy half (1965–1970)	40.0% silver—400 fine
U.S. silver War Nickel (1942–1945)	35.0% silver—350 fine

Even if people are aware of the fine value of certain coins, they are unlikely to be able to relate that value to the spot metal prices and the prices a gatherer offers.

8

Assaying and Refining

No segment of the precious-metals industry is more important than the refiner. No research you might do in relation to proposed precious-metals dealings is more important than finding an established refinery. If you plan on taking possession of your refined material, as opposed to selling the scrap metal outright to the refiner, you should deal only with a refinery whose bullion bars are recognized in gold-trading centers. If you had bars produced by an unrecognized refiner, you might be faced with melt-and-assay charges when you go to sell them.

In my research for this book, I checked into many refining operations. Some, like Engelhard, and Handy and Harman, are well-known firms. Their integrity is considered by most people as unimpeachable. Engelhard does not ordinarily deal with individuals; their business is almost entirely with industrial accounts. There are smaller firms like Westbury Alloys who come highly recommended and have much to offer the precious-metals market. There are also firms that advertise as refiners, but actually are no more than gatherers themselves.

There is even a place for the smallest refinery if it operates honestly. These businesses can buy your precious metals outright, or on a melt-and-assay deal, so you have no worry about possessing bullion with an unrecognized name. Of course, even the largest refiners will buy your precious metals on a melt-and-assay deal, as well as produce the bullion for your personal ownership.

Many refiners offer different services, and not all refiners will handle every form of precious metal. Some are specialists, and it requires research to discover the refiner who best meets your needs.

A good starting point would be any of the firms mentioned at the end of this section.

The prices and figures given here are not to be considered a standard for refiners. They are average figures that were taken at the time this book was written, and could change. They are used here to establish a reasonable price schedule to help you determine the fairness of a deal offered by a gatherer. These figures were taken from a firm that deals only on a melt-and-assay policy. After melting and assaying, they will either purchase the refined precious metal or return the refined precious metal to you, charging only for the assaying and refining (with an adjustment to cover refinery loss).

The cost of refining differs for gold and silver. Gold currently has a minimum charge of $275 (on average), or $0.60 per ounce, whichever is greater. This means that any amount up to 458.33 oz. (275 ÷ 0.60)—whether 1 or 300 oz.—costs $275 to refine. For amounts of gold greater than 458.33 oz., the refining charge of $0.60 per ounce applies. If you had 550 oz. of gold, the assaying and refining charge would be $330.

Refining silver alloy is figured differently. A minimum charge of $275 is made for any amount up to 1,500 oz. From 1,500 oz. up to 4,000 oz., the price is $0.18 per each additional ounce. Therefore, 1,510 oz. costs $275 + $11.80 = $276.80 to refine. Each additional ounce over 4,000 oz. costs $0.16 per ounce to refine.

Adjustments are also made to take into account the small amount of precious metal that is lost during refining. Gold (with a melting point of 1064.43°Centigrade) volatilizes into a vapor at temperatures higher than its melting point. Simply put, gold evaporates when you boil it. Such volatilization is slight near the melting point, but with an increase of 187°C, for example, the hourly loss is about 2.6 parts per 1000. Refiners systematically clean out their exhaust systems to reclaim as much of this lost gold as possible.

A good rule of thumb for both gold and silver is to calculate the refining loss at about 2 percent per ounce of fine precious metal. To calculate refining loss, multiply the fine-gold-content figure by 2 percent. Multiply that result by the spot gold price; this gives you the value of the 2 percent loss. Next, you can multiply the fine-gold-content figure by $0.60, to give you the refinery fee. Adding the two gives the total charge. For example, we will use 1 oz. fine-gold content at a spot gold price of $580:

1.	1.00	1 oz. fine gold in the alloy
	×.02	% of loss
	.02	oz. refining loss
2.	$580	spot gold price
	×.02	refining loss
	$11.60	the cost of refining loss at a spot price of $580
3.	1.00	1 oz. of fine gold in the alloy
	×$0.60	cost of refining 1 oz. of gold
	$0.60	cost of refining 1 oz. of gold
4.	$11.60	refining loss
	+$ 0.60	refining cost
	$12.20	total charges for refining 1 oz. of fine gold from an alloy at a spot gold price of $580

Any time refinery loss and refining charges are discussed in this book, they are based on the gatherer paying the minimum refining charge.

At a time of "crunch," or when there is a heavy demand for refining services, the policies of melt-and-assay firms can tie up your cash for extended periods. The period of intense interest in scrap metal of the recent past saw backlogs of six weeks and more. There were ways some of the melt-and-assay refiners made this easier on established customers in the beginning, but after a while, even preferred customers had serious cash problems.

"Melt and assay, then pay" is not the policy of all refiners. Many small operators will pay on the spot when you bring in your precious metals. Some buy both silver and gold, while others buy only gold.

Example: One such refiner, when the spot price for gold was $515.70 per ounce, was paying outright $13.15 per pennyweight (dwt.) to buy 14K gold without first melting and assaying it. To determine what price this smelter was paying for the fine gold in the 14K alloy, the $13.15 must first be multiplied by 20 (20 dwt. equals 1 oz.): $13.15 × 20 = $263 per oz. of 14K gold. Fourteen-karat gold is 583 fine, 13½K gold is 563 fine. Fourteen-karat gold, 583 fine, has .583 oz. fine gold; 13½K gold, 563 fine, has .563 oz. fine gold. Since the spot gold price was $515.70 per ounce, the market value of 1 oz. of 14K gold was $300.65 ($515.70 × .583). The value of the 13½K gold was $290.34 ($515.70 × .563).

Thus at the $515.70 spot gold price, and based on the refiner's offer of $13.15 per pennyweight, the refiner makes a profit of $37.65 per ounce of 14K gold ($300.65 − $263.00); on 13½K gold the refiner's profit is $27.34 ($290.34 − 263.00) per ounce.

With an arrangement as in the above example, you do not have the cash-flow problem related to a melt-and-assay deal, in which you must wait until the alloy is melted and refined. You know at the outset the price you will receive, and you are paid on delivery. However, you likely will get a better return on your gold scrap at a melt-and-assay operation, if you don't experience a long wait with an intervening drop in the spot gold price. There is one way to beat the wait at a melt-and-assay refiner: If the dealer previously offered a settlement at a price set to disregard spot price fluctuations, you can be assured of the amount of payment. You should check with the refiners if you are interested in such a deal.

The names and addresses of some refiners are:

Westbury Alloys Corp.
750 Shames Drive
Westbury, N.Y. 11590

Handy and Harman
850 Third Avenue
New York, N.Y. 10022*

Frank Mossberg Drive
Attleboro, Ma. 02703

4140 Gibson Road
El Monte, Ca. 91731

141 John Street
Ontario, Canada M5V2E4

* Offices only

American Chemical and Refining Co.
Waterbury, Ct. 06720

Simmons Refining Co.
4105 West Chicago Ave.
Chicago, Il. 60651

Engelhard Industries Division†
Engelhard Minerals & Chemicals Corp.
429 Delancy Street
Newark, N.J. 07105

9

The Acid Test for Precious Metals

A refiner melts and assays precious metals to determine their purity. However, jewelers, coin dealers, gatherers, and other persons interested in determining the purity of silver and gold cannot wait to send the items in question to the refinery.

A relatively simple method of testing such items has been in use for many years. The acid test requires a testing stone, nitric- and hydrochloric-acid solutions with an applicator, a metal file, and (for close accuracy) testing needles of specific fineness.

The first step is to verify that the item is at least karat gold, not just gold plate on a base metal. A drop of nitric acid is applied to a slight notch filed into the piece. If the reaction to the acid is a bright green, the piece is gold-plated copper. Hardly any reaction at all, no more than a soft sizzling, means that the piece is an inferior gold alloy of less than ten karats. Gold plate on silver base is shown by a pinkish-cream reaction. If there is no reaction at all, the piece must be tested further.

From this point on, the test can either be precise or merely a good guess, depending on the ability of the tester and the equipment used. Many persons use only the testing stone and acid for the determination of purity but this is not nearly as accurate as using the

† Mainly for industrial accounts

testing needles. Precision testing with the needles depends upon the close attention of the tester. The piece in question is rubbed across the test stone to leave a mark. Next, the needle that has the same designated karat or fineness as the piece is rubbed across the stone beside the first mark. A solution that is three parts hydrochloric acid to one part nitric acid *(aqua regia)* is run across the two marks with the applicator. If the mark from the piece in question reacts before the mark from the needle, the piece has a lower karat than the karat stamp on the piece. A needle with the next lowest karat designation must be used in another rubbing test. This process must be repeated until the reaction between the two marks is the same. The needle with the same reaction time as the piece shows the actual fineness of the piece. The karat designation placed on the piece by the manufacturer can be disregarded.

Silver testing is not as complicated. To test whether the piece is silver or silver plate, a notch is filed in the piece and nitric acid is applied. If the acid produces a cloudy cream coloration, the piece is sterling. If the nitric acid causes the notch to turn green, the piece is silver plate on a base metal.

10

Deviations in Gold and Silver Manufacturing

It is possible that a gold (or supposedly gold) piece to be tested will not have a karat designation stamped upon it. Prior to 1906, jewelry was not regulated by the government, and many manufacturers or producers of jewelry put neither a purity indication nor their own identifying stamp on their work. From 1906 through 1961, although there were changes made in the National Gold and Silver Stamping Act, a producer had only to mark the karat designation on the manufactured article. Since 1961, the manufacturer also had to put his trademark on his work.

The 1906 Act was lax in several ways. It allowed manufactured gold items (other than watchcases and flatware) to be one-half karat less than the mark stamped on the piece if the piece had no solder in it, and a full karat less if the piece contained solder. Watchcases and flatware were required to be no more than three one-thousandths off their mark. Sterling and coin silver were allowed to be up to ten one-thousandths off their mark if they contained solder, and four one-thousandths off if there was no solder in the work.

To make this deviation from the stampings more understandable, the following chart shows the figures in fineness (parts of 1000).

Deviations in Fineness of Karat Gold as Allowed by United States Law (Stamping Act of June 13, 1907)

	STAMPING	DIVERGENCE ON SOLDERED PIECES	DIVERGENCE ON NONSOLDERED PIECES
Gold— Jewelry (24K)	999 fine	957 fine	978 fine
Gold— Watchcase and Flatware (24K)	999 fine	996 fine	—
Sterling Silver	925 fine	915 fine	921 fine
Coin Silver	900 fine	890 fine	897 fine

Of course, coin silver at 900 fine and sterling silver at 925 fine are the only two fineness designations commonly found on silver. Gold is used in fineness from 10 karat (417 fine) to 24 karat (999 fine). The deviation on jewelry gold is always twenty-one one-thousandth on nonsoldered, and forty-one one-thousandth on soldered, regardless of the karats. The deviation on watchcase and flatware gold is always three one-thousandths, regardless of the karats.

On October 1, 1981, these allowances will change for gold used in jewelry. Watchcase- and flatware-gold deviations, as well as coin and sterling-silver deviations, will remain the same. Jewelry gold will fall under these new allowances:

New Deviation Allowances (as of 10/1/81) on Jewelry Gold

	STAMPING	DIVERGENCE ON SOLDERED PIECES	DIVERGENCE ON NONSOLDERED PIECES
Jewelry Gold (24K)	999 fine	992 fine	996 fine

Again, for other karat-gold designations, the deviation will always be seven one-thousandths on soldered, and three one-thousandths on nonsoldered, gold.

This new law, tightening the amount a manufacturer can deviate from his marked designation of purity, is an improvement over the original law, but the fact is alloyers had the ability to produce "plumb" gold when the first Stamping Act was passed in 1906. Plumb gold is gold alloy accurate in purity to the mark stamped on the finished product. A bracelet marked 14K which, in fact, is actually only 13½K, is a deviation from the mark. Fourteen-karat gold that is a full 14K is plumb gold. Deviations from the mark obviously affect the value of the gold item. (For more on this, see "Karat Gold and Plumb Gold.")

Many jewelry, flatware, and other precious-metals manufacturers have *always* used plumb gold in all their products, for example, Gorham Textron, which has furnished information and photographs that are featured in the jewelry and vermeil sections. The use of plumb gold and plumb silver is an accepted policy with Gorham, as it is with many other fine manufacturers. As a matter of fact, some manufacturers are insulted if they are even asked if their products are plumb metals. The use of plumb gold and silver is as much a part of their businesses as their names.

The Stamping Act of 1906 was the start of a slow transition of American precious-metals manufacturing from basically self-policing to a fairly well regulated field. In the old days, before the Act went into effect, a manufacturer could turn out a product with no purity designation, and no mark of identification as to who made the item. Many pieces of jewelry sold as 14K gold were only found out to be less pure at a later date. With no identifying mark from the maker, there was no one to call to account for the dishonesty.

Further, many misleading terms and names were made illegal for use on precious metals; for instance, *United States Assay* or any words, phrases, or devices to give the impression that the United States government had certified the purity of the product. It became illegal to stamp or mark silver- or gold-plated products with words such as *sterling, coin,* or *gold* without also indicating that the product was only surface-covered with a precious metal.

One of the most misleading terms used in precious metals is *solid gold.* The Federal Trade Commission has ruled that ". . . any article that does not have a hollow center and has a fineness of ten karats or higher . . ." can be called "solid gold." Many people confuse solid gold with pure gold. Solid gold can be as low as 10 karats (9K with deviation), while pure gold is 24 karats.

Part Two

Your Gold

11

Gold

Man has used this soft metal for more than 8,000 years. In all those years, the earth's most accessible gold ore has been found, and it is now necessary to bring to the surface an average of three *tons* of ore to extract one *ounce* of gold. All the gold extracted since the beginning of time would cover a football field only to a depth of approximately 42 inches. Experts maintain that there is only enough undiscovered gold left to satisfy our market needs for possibly another twenty years.

What are those needs? And what actually has happened to that waist-high football-field-size accumulation of gold? The United States government holds the largest reserve, followed by private French citizens, who own more gold collectively than the International Monetary Fund, which is third in line. Private citizens of India rate fourth in holdings, with the West German government a close fifth. In the short time American citizens have been permitted to possess gold, they have put themselves in sixth place in the gold sweepstakes. The French government is in seventh position, holding little more than half as much as its private citizens. The Soviet Union, with perhaps the greatest gold supply still in the ground, holds eighth place with its banked supply; the Swiss and Italian governments run a close race for ninth and tenth place. All the remaining gold held by governments other than those listed accounts for less than the combined holdings of the United States and French governments. Finally, the holdings of other individuals, widely distributed around the world, is the rough equivalent of the holdings of French citizens.

That football field of gold would weigh two-and-one-half-billion troy ounces. There is a great demand for gold in its many forms—bullion, jewelry, and coins, to name a few. How is that demand met?

The supply of gold is not static. An investor might retrench his holdings or see a chance to make a profit on his original investment; the Soviet Union might use some of its reserve to buy farm machinery; the boat people of Vietnam sold large collections to finance their escape. A prime example of the continual movement of gold is when the price of gold encouraged millions of Americans to search through drawers and to literally take the jewelry off their hands to sell to gatherers. Of the two-and-one-half-billion ounces of gold now held throughout the world, over one-half billion ounces were on the market in one form or another last year.

To understand gold, you must put it in its proper perspective. Some experts believe it to be the soundest instrument to back up world trade, while others feel gold should be divorced from international monetary dealings.

It is said that gold has little intrinsic value. If that is true, by the same reasoning you can so classify aluminum, wood, rubber, copper, oil or any material man has put to use. The value of any product is found in the need and desire of man to utilize that product. The practical side of man would not perpetuate a useless material for centuries.

Far from being impractical, gold is a very useful material. We are most familiar with gold as jewelry or coins, or possibly we envision the piles of ingots stacked neatly at Fort Knox. We don't see the multitude of uses gold is put to in electronics. You might be amazed at the number of products in your home that contain gold, for example, television sets, calculators, and microwave ovens are just a few. Every time you make a phone call, gold helps make your connection. The electrical conductivity of gold is outstanding, but possibly even more important is the exceptional resistance gold has to tarnishing and oxidation. Gold is far from being a "vain" metal good only for viewing and hoarding. Its uses vary about as much as any other metal, new uses being limited only by the expense of the metal.

The high cost of gold, subject to wide fluctuations, is not totally unrealistic. Eight hundred dollars an ounce might be considered unrealistic but the long-prevailing price of $35 an ounce was equally unrealistic. Gold is, and always has been, a scarce metal. If it wasn't, and the supply exceeded demand, the price would be no higher than that of any other common metal. After all, for some time, aluminum commanded exorbitant prices in the early days of its discovery. Iron, copper, tin, all the nonprecious metals are mined with regularity. There are occasional shortages, but only until production can be increased to meet greater demands. Production of gold cannot be easily increased. The days of the California Gold Rush—the days of readily prospected nuggets of gold—are gone.

The great value of gold assures that it is never knowingly thrown away. You might toss out your old lawn mower, with its stamped- or cast-steel body and parts, but your old gold ring is sold for scrap. That ring will be used and reused through countless applications. It is believed that one of the reasons no gold artifacts are found dating prior to 8,000 B.C. is that the following civilizations melted down and reworked previous artisans' works. We know that many succeeding emperors and kings melted down their predecessors' coins, minting currency bearing their own image. It is improbable, but not inconceivable, that the gold in the new ring bought today could contain some gold from a trinket worn by Cleopatra.

We see ornate decorative gold leaf on public buildings, but gold has a more practical use in today's structures. Very often modern buildings, with designs incorporating large surface areas of glass, are saving energy with insulating glass coated with gold. Ordinary glass, despite the use of thermal sandwiches of glass with captive air spaces in between, transmits the sun's rays with little resistance during the summer. During heating seasons, much energy is lost through the windows. A thin coating of gold on the inside of these insulated windows can reflect over ninety percent of the sun's infrared heat rays, while passing enough light to permit sufficient lighting inside the building. This reflectivity also reduces energy costs by saving on air-conditioning operations.

A building five stories high and one hundred feet long, with an all-glass face, would use no more gold than found in five Krugerrands. The savings in fuel for heating and air-conditioning could more than pay that cost in a matter of the first few years!

Gold's malleability (or workability) has been the chief factor in its long use in jewelry and adornment. Ancient artisans found they could beat gold to a thinness of one-quarter-millionth inch. An ounce of pure gold so worked could cover an area of 100 square feet. That thin one-ounce sheet could be rolled up, reworked, and drawn into a wire fifty miles long. If you accumulated and formed a cube of gold twelve inches by twelve inches by twelve inches, it would weigh over a thousand pounds.

However, the workability of gold is also one of its disadvantages. It is so soft that in an unalloyed state as a coin or jewelry, it would quickly wear out in extended use. Because of this, alloying has long been an art. The use of copper, zinc, nickel, and other metals creates harder alloys, as well as different colors and tones.

For years Americans were forbidden to own gold in many forms, and many thought we were alone among the citizens of the world in such a restriction. Actually, many governments do not allow their people to own gold, or, at least, limit ownership. Such limitations do not necessarily prevent the people from possessing gold, however. Gold has traditionally been considered safe in "hard" times. It is internationally accepted as collateral, while a nation's paper money can become virtually worthless. Some countries don't allow possession of bullion or coins of gold, but allow ownership of gold jewelry. The rules may differ from country to country, but the attitude of the people rarely changes: If they can't own gold legally, they hold it secretly. As a matter of fact, the clandestine ownership of gold among citizens of restricted countries probably helps account for the vast amount of gold that is smuggled each year. It is estimated that one-third of the gold refined each year passes into the hands of smuggling operations.

It's known that the ownership of gold saved countless lives when families had to flee Europe before and during World War II. Many Vietnamese refugees also found gold to be their salvation in buying freedom.

12

Karat Designation in Relation to Fineness

Except for coins, gold is usually designated by *karat.* One karat is one twenty-fourth part gold in the item. Karat must be changed to fineness to calculate the value of your gold. The regular conversion for karat gold follows. To consider the deviations below actual karat marks or stamps on a gold item, see the chart found under the section "Karat Gold and Plumb Gold" on page 46.

Converting Karat Gold to Fineness
24-karat gold is 99.9% pure—999 fine
22-karat gold is 91.7% pure—917 fine
18-karat gold is 75.0% pure—750 fine
16-karat gold is 66.7% pure—667 fine
14-karat gold is 58.3% pure—583 fine
12-karat gold is 50.0% pure—500 fine
10-karat gold is 41.7% pure—417 fine

Twenty-four-karat gold is pure gold (also called fine gold), while 14-karat gold is fourteen-twenty-fourths of pure gold, with ten-twenty-fourths being an alloy metal, such as copper, zinc, nickel, or even silver.

Twenty-four-karat gold is considered pure gold. Pure gold is the basis for the spot gold metal price found in newspapers and other publications. Many precious-metals dealers have computers tied into the world gold-and-silver trading centers, and they can obtain the price current at that moment. The price can change several times daily, and at any one moment the price can differ at any of the widespread trading centers.

Although 24-karat gold (999 fine) is called "pure," pure is a relative word. It is not possible to refine completely pure gold (or silver). The difference of one one-thousandth to complete purity at a spot gold price of, say, $500 an ounce, is only fifty cents.

"Four nines" (9999 fine) is usually the purest form that gold bullion comes in. It is generally shown on the bullion bar as *999.9*. If the positioning of the decimal point, or the lack of it, seems confusing, you can effectively forget about it. "Four nines"—9999; 999.9; .9999; 99.99; 99.99 percent (at least in gold parlance)—all mean the same thing. One might be the purity in fineness, another the percent of purity, another the purity carried to the nearest thousandth, each still referring to an identical degree of purity.

Both 999 and 999.9 fine have been used in this reference to pure gold. In various books and publications you will find reference to purity of gold by other, more accurate degrees of fineness. The following list gives the commonly used figures of greater accuracy past 999 fine.

Degrees of Purity of Fine Gold

999	= .999000 =	999	parts of 1000
9995	= .999500 =	999.5	parts of 1000
9999	= .999900 =	999.9	parts of 1000
99995	= .999950 =	999.95	parts of 1000
99999	= .999990 =	999.99	parts of 1000
999995	= .999995 =	999.995	parts of 1000

You could say that all the above degrees of fineness are 24K gold. Jewelers start their alloys with .9995 gold, where bullion is generally sold as pure at .9999.

If you look up the daily spot gold price in a publication, you will usually find it listed as "unfabricated" or "fabricated." The price you are interested in is the unfabricated price. The fabricated price is for specially refined gold used for industrial purposes. The greater degree of purity required by some industries calls for more concentrated refining efforts, and consequently, higher cost. As a matter of fact, the fabricated price quoted is not the highest-priced gold available. Some specialized industries require gold fabricated to such exacting standards that the cost would intimidate the average investor.

Fineness Range of 24-Karat Gold

FINENESS	USE
.999000 .999500—jewelry .999900—bullion	unfabricated
.999950 .999990 .999995	fabricated (industrial)

13

Scrap Gold

Scrap gold is the general name for jewelry; watches; dental work; damaged gold coins; and a vast array of shapes, sizes, and items.

It is probably difficult for anyone to think of a material with gold's value as scrap. Americans are familiar with the mad Great Precious Metals Scrap Drive of 1979–1980, when the ever-spiraling price of gold turned even the most staid individuals into attic- and drawer-scroungers, searching for *any* item with silver or gold content. Their discoveries, while not always productive, were often at least humorous. Some of the more commonly repeated stories centered around the public's general lack of knowledge about silver and gold. The anecdotes that follow show these touches of humor. Harvey Ravitch of Westbury Alloys was kind enough to share some of his experiences with me.

Telephone rings. "Good morning, can I help you?"

"Yes, I wonder if you can give me the formula for extracting gold from seawater?"

"Well, I'm afraid there really is no efficient formula for such . . ."

Interrupting: "No, I know there *is* a formula. I'm willing to pay."

Telephone rings. "Good morning, can I help you?"

"I hope so, I have a large quantity of gold I'd like refined."

"Certainly sir, can you give me an idea of what quantity and form it's in?"

"It weighs about a hundred pounds and it's in the form of a rock in my back yard."

I repeat these stories to point out that the public is basically uninformed about silver and gold. These incidents were not uncommon; they reflect the public's need for concise information.

People who feel that this recent near-hysteria to find and sell precious metals was a rare occurrence, a phenomenon unknown in the past, are wrong. It was probably the most public such happening, an extravaganza encompassing more nations than ever before, but such dealings have been going on since silver and gold went from being possessed solely by kings to being the ultimate prize of the masses.

If it wasn't for scrap gold being turned in, the market for all finished gold products—bullion, jewelry, coins, etc.—would suffer. The constant rotation of gold holdings is needed to meet demand. If all gold held by individuals, companies, banks, nations, and others was to suddenly be held off the market, and only newly mined gold used to try to satisfy the demand, the price would soar to new heights. Such a condition, if it *was* precipitated, would not last for long. As soon as the price of gold rose dramatically, people would sell, creating an increased supply. This would effectively drop the price of gold.

When scrap is initially traded, the gatherer usually determines the karat by acid-testing the item. Some dealers are fairly exacting, for the acid test can pin fineness down within at least a half karat; but many are not too concerned about being very close to the karat mark. At the price they pay for scrap gold, they haven't much to worry about; especially now that the crunch is gone from the refiners.

A gatherer can pay his 50 percent to 200, or in some cases as much as 800, percent under the true value of the gold content, take his purchases up to the refinery once a week, and live comfortably.

Scrap gold is different from scrap silver in several ways, other than the actual physical characteristics of the two metals. First, when dealing with scrap silver, you know that, if the item isn't silver plate, it is probably 925 sterling (considering the possibility of the allowable fineness deviation). You can test a batch of scrap sterling for plate, then weigh the pieces together because they are the same fineness. Scrap gold is different. You have 10K, 12K, 14K, 18K, 22K, and even 24K gold (damaged gold coins might be 24K). Of course, most American jewelry is 14K (allowing for deviation), but at its high cost per ounce, greater care is generally taken in evaluating the fineness of gold scrap than silver scrap.

Many attempts are made to hoodwink the gatherer. Sometimes crude efforts to scratch 14K on a piece of gold plate are made, while more sophisticated engravings and stampings are also counterfeited. The buyer isn't the one often taken advantage of, however.

Another form of scrap gold that should be mentioned specifically here is scrap gold coins. First, you must understand specifically what gold coins should be considered as scrap. Certainly, numismatic coins are not considered scrap. Such coins, quite often, have been in circulation and possibly bear nicks, scratches, and other marks, and sometimes are even bent. They also can exhibit signs of wear. Depending on the scarcity of the coin, such defacing marks may detract from the numismatic value, but the value could still be greater than the intrinsic gold content. Some numismatic coins are in uncirculated or "proof" condition, and they are even more valuable.

Bullion coins are not usually issued for circulation, but there are original-issue coins that might have been in circulation and still might be classified (at least marginally, if not fully) as bullion coins. Or, they might be official government restrikes such as the Austrian 100 corona. Then there are coins such as the Canadian Maple Leaf and the Krugerrand that are original-issue gold coins specifically for bullion sales. Whatever the qualification of the bullion coin, it should be kept in mint condition. Many freshly issued gold-bullion coins are packaged in clear plastic covers. They should be kept in that cover.

When a bullion coin, a gold coin with no numismatic value, becomes damaged, and the dealer determines the quality of the coin is diminished too much, it becomes necessary to treat that coin as scrap. Do not accept the word of a person or a dealer of whose integrity you are not sure that a gold coin is good only as scrap. Even a friend with good intentions and a mistaken belief that he knows what he is talking about should not be accepted as an expert on gold coins.

14

Some Calculations for Scrap Gold

Dealing with a precious-metals buyer requires no special skills. You do not have to rely upon his expertise and knowledge in the transaction. If you have a copy of this book, a pocket calculator, and a pad and pencil, you can enter the deal with as much basic knowledge as he has. Of course, his experience in such transactions will certainly outweigh yours, but that is not necessarily a disadvantage.

To determine the actual fine-gold weight value of your piece, first write the daily spot gold price (as it appears in a publication) at the top of your pad. Write the fineness figure of the gold piece on the pad also. To get the correct figure, look on the preceding chart (page 37). The fineness figure you want is the one that corresponds with the karat mark found on your gold piece.

The dealer's scale will be graduated in grams, troy ounces, pennyweights, or grains. If it is graduated in any unit other than ounces, you will have to convert the weight shown on the scale to ounces, because the spot gold price is given in ounces. The method for conversions follows.

For example, we will say that you have a gold bracelet marked *14K* (583 fine).

If the scale, using grams, shows a weight of 307 grams, divide 307 g. by 31.104 (the number of grams in one ounce): We get 9.87 oz.

If the scale, using pennyweights, shows a weight of 197.4 pennyweights, divide the 197.4 dwt. by 20 (the number of pennyweights in an ounce): We get 9.87 oz.

If the scale, using grains, shows a weight of 4737.6 gr., divide the 4737.6 gr. by 480 (the number of grains in an ounce): We get 9.87 oz. You now have the weight in ounces of the gold alloy of the item in question. You must now determine the fine-gold-content weight. Simply multiply the ounce weight of the alloy (the weight you have just established) by the fineness figure. *You must always place a decimal point in front of the fineness figure at this stage of calculations.* The result is the actual fine-gold weight of your gold item: 9.87 oz. × 0.583 fine = 5.75421 oz. fine gold.

To find the actual value of the fine gold as based on the daily spot price, multiply the result (5.75421 oz.) by the spot gold price. For example, we will say the spot price is $700 an oz.; therefore, 5.75421 oz. × $700 = $4,027.95. The full value of the fine gold of your bracelet is $4,027.95.

In any trading of scrap precious metals, the buyer must figure into the purchase, and deduct, the cost of assaying and refining. Further, there is up to a 2 percent loss of fine gold (or silver) during refining, and this also must be considered. (See the section titled "Assaying and Refining" for more details.)

Thus, in the final calculation of the actual value of the fine-gold content of the bracelet weighing 9.87 oz., the refining and assaying cost to deduct would be $5.92 ($0.60 per ounce × 9.87 oz.), and the 2 percent refining loss to deduct would be $80.56, for a total deduction of \$86.48. Thus we have \$4,027.95 (full value of the fine-gold content) − \$86.48 = \$3,941.47.

The actual value of your 14K-gold bracelet weighing 9.87 oz., with a fine-gold weight of 5.75421 oz., deducting average refinery charges and losses, is \$3,941.47 (at a spot gold price of \$700 an oz.).

The sheet of your pad, with the calculation for the 9.87-oz. bracelet, should look like the following (the spot gold price, along with your calculations for the refinery loss/assay and refinery charges, and the fineness of the bracelet should be done at home before going to the dealer):

Spot Gold Prices	\$700.
Bracelet fineness	.583
refining loss, assaying refining costs	\$86.48

weight - 307. grams

307. grams = 9.87 oz.

9.87 oz. × .583 fine = 5.75421 oz. fine weight

5.75421 oz. × \$700. = \$4027.95

− 86.48

\$3941.47 Value of fine gold

Dealer's offer: \$

Copy all these notations onto your pad, even though you are using a calculator, so you will have them for immediate reference and later record.

The value of $3,941.43 reflects the deductions for an average refining charge and loss. Such charges are subject to change, and different refiners might have different price schedules. The figures given in this book are for your use in determining a relatively accurate figure. For total accuracy, you would have to contact a reputable refiner for the latest figures. (A short list of such refiners is found in the section on refineries, on page 25.)

Of course, if you are selling to a gatherer, you will not be offered anywhere near $3,941.43 for the bracelet. Many such buyers have been paying 50 percent to 200 percent, and some as much as 800 percent, under the actual cash value of the fine gold. *Knowing the actual fine-gold value puts you in a better negotiating position.* Many dealers, however, are so used to the exorbitant profits they have been making that they may not be interested in dealing at a correct rate. It will not be until the majority of the trading public is properly educated in the value of their gold scrap, that the gatherers will start dealing more equitably.

One alternative is simple. I believe that there is a sufficient number of people in a given area to form an association of sellers. They could pool their various gold and silver items, accumulating enough volume to qualify for the minimum refining charges (if the refiner they work with has such charges). The selected head of the association could act as an agent in dealing with the refiner. For his services, he could be paid a specific percentage of the refined price of the items from each member. Under this system, dispensing with the middleman, the members would be realizing a return based on the actual value of their precious-metals scrap, minus the assay-and-refining charge, of course. If that seems like too much trouble, remember, many gatherers are realizing up to (and sometimes over) a 200 percent profit on their purchases.

If you form an association, I suggest you check out any refiner at the plant before concluding a deal. The established, reputable refiners will welcome such action, and the questionable ones will become obvious. Many so-called refiners I checked into were no more than gatherers themselves, hiding behind the title "refiner."

If you have a gold coin that is definitely determined to be scrap, it should be figured basically as the bracelet example, with one major exception. Some gold coins, such as the Maple Leaf and Krugerrand, have the specific fine-gold content marked on them. Most coins, however, must be researched to find their fine-gold content. This book lists all the popular bullion coins, along with all the gold coins minted as legal tender in the year 1979 (the latest year such data is available as this book goes to press). Almost any gold coin can be found there. If you have a coin older than "modern coinage," it most likely has numismatic value (as do many of the coins listed under "Worldwide Gold Coinage"). For an example of the method of evaluating a damaged gold bullion coin, a Mexican 50 Peso will be used.

The Mexican 50 Peso is listed in this book's charts as having a total weight of 1.3396 oz. and a fine-gold content of 1.2057 oz. The fineness is 900. When the fineness is a known factor, such as in listed gold coins, there is no need to calculate that fineness as in other scrap gold. It is only necessary to multiply the ounce figure of the fine gold by the spot gold price. Say the spot price is $580 per ounce, the figuring would be: 1.2057 oz. × $580 spot price = $699.306. The full value of the fine gold of the damaged 50 Peso would be $699.31. Deducting the refinery loss and the assaying-and-refining charge, which total $14.71, the damaged bullion coin has a gold sale value of $684.60. The list of United States gold coins gives pertinent information, but to determine the numismatic value of any of the coins, you should consult an authority.

15

Karat Gold and Plumb Gold

The term *karat gold* refers to gold of at least 10K, or ten-twenty-fourths pure gold. A 1959 ruling by the Federal Trade Commission requires the minimum of 10K purity or the alloy cannot be called gold.

As is so common with the government agencies and regulations, that 10K-minimum-purity rule is subverted by another government law that allows deviation from the stamped markings on precious-metals products (also read the section titled "Deviations in Silver and Gold Manufacturing") until October 1, 1981.

To show the actual deviations in figures, so that you can evaluate gold based on its true fineness, the chart shows the actual finenesses of gold deviating ½K and 1K from the true fineness of karat designations.

One-Half and Full-Karat Deviations of Karat Gold

TRUE	DEVIATION OF ½K	DEVIATION OF 1K
24K gold, 999 fine	23½K gold, 979 fine	23K gold, 958 fine
22K gold, 917 fine	21½K gold, 896 fine	21K gold, 875 fine
18K gold, 750 fine	17½K gold, 729 fine	17K gold, 708 fine
16K gold, 667 fine	15½K gold, 646 fine	15K gold, 625 fine
14K gold, 583 fine	13½K gold, 563 fine	13K gold, 542 fine
12K gold, 500 fine	11½K gold, 479 fine	11K gold, 458 fine
10K gold, 417 fine	9½K gold, 396 fine	9K gold, 375 fine

The first column is *plumb gold,* that is true gold in purity of marking. Plumb gold is any gold alloy that is true to its mark or stamp. In other words, 14K plumb gold *is* 14K, not 13½K. Understand though, plumb refers to accuracy in marking only. Conceivably, a manufacturer could mark his product 13½K, and if it was 13½K, that would be plumb gold. As long as the actual content of the product is as pure as indicated by the marking, that is plumb gold. Of course, it is highly doubtful anyone is going to use such an unusual mark.

To understand how the chart relates to the value of an actual gold piece, the following example is given: Say the spot price for gold is $700 an ounce and you have a bracelet that weighs one full ounce. It is marked 14K, but tests at 13½K (or 13K).

Using the formula we have demonstrated elsewhere (spot price × fineness), we find that the actual gold value of the bracelet, at a spot gold price of $700, would vary: 14K = $408.10; 13½K = $394.10; 13K = $379.40.

Understand that not all manufacturers take advantage of this bonus the government has allowed them for many years. The only way you can find out if your jewelry item is plumb gold is to ask. Some manufacturers will gladly tell you, some might not reply, and you might be told that you are buying the piece as a work of art, and the specific fineness doesn't matter.

You must also understand that manufacturers who stringently adhere to the plumb-gold policy can have jewelry with a fineness less than the marking when soldering is involved. This is only because the solder work is intricate, but they have not started out with one-half karat less in the actual gold used. Any deviation from the full purity of the karat designation is restricted to exactly what is necessary to produce quality workmanship.

16

Other Gold Designations

Gold Filled, Rolled Gold Plate, Gold Overlay, Gold Plate, Gold Electroplate, Heavy Gold Electroplate, Gold Wash(ed), Gold Flash(ed)

All of these words or terms—for gold surfaces on a base metal—are restricted in use with products manufactured or sold in the United States, along with other markings of abbreviations or initials that would be used to indicate items made of precious metals.

If the marking *gold* alone is used, with no other word or number to indicate form or fineness, the product must be 24K gold. An item cannot be marked *9K gold* because the law requires gold to be at least 10K before it can be classified as gold (see the section "Deviations in Gold and Silver Manufacturing" for exceptions).

Gold Filled. A product that has a layer of (at least) 10K gold mechanically bonded by heat and pressure to a base metal. The karat gold surface must comprise at least one-twentieth, or 5 percent, of the total weight of the metals. Such a product can be marked by the karat designation of the gold layer, such as, *14K gold filled.* It is not necessary to indicate the one-twentieth gold weight because that is the known standard. G.F. is the abbreviation for gold filled.

Rolled Gold Plate. The same basic requirements as for gold filled, that is, at least 10K gold layered on a base metal, but the proportion of karat gold to the base metal can be less than the one twentieth of the total weight if the specific percentage of karat gold is indicated clearly, such as, *1/40 14K Rolled Gold Plate* or *1/40 14K R.G.P.*

Gold Overlay, Gold Plate. Like *rolled gold plate,* these terms can be used as markings on gold-layered products without a fractional prefix only if the gold weight is at least one-twentieth of the total weight of the metals. If rolled gold plate, gold overlay, gold plate, or any of their abbreviations is used as a marking on a manufactured product that has less than one-twentieth of karat gold in the total weight, the fractional percentage *must* be noted on the marking. For example, 12K gold overlay with a one-fortieth proportion of karat gold to the total weight would be marked *1/40 12K Gold Overlay.*

All markings and stampings on manufactured items require the karat designation to be included, with the exception of *pure gold,* which must use 24K gold (and no karat designation is then needed). All markings must be equal in size so as to prevent deception.

Heavy Gold Electroplate. When a thinner layer of karat gold is desired on the base metal, an electrolytic process is used. Gold filled is bonded by heat and pressure, while gold electroplate is applied by use of an electric current while the product is suspended in a liquid solution. The law requires a minimum thickness of 0.0001 inch karat gold on heavy gold electroplate.

Gold Electroplate. The same electrolytic process is used, but the karat-gold thickness of the plating can be a minimum of 0.00007 inch. If electroplated karat gold does not meet the minimum 0.0007-inch standard, it can be called gold flash(ed) or gold wash(ed).

From the descriptions of the various designations for gold surface-covered base metals, it is obvious that gold flash, gold wash, gold electroplate (and silverplate), and heavy gold electroplate are not ideally suited for trading as scrap. With no more than one-ten-thousandth-inch thickness of precious metal over a base metal of perhaps a sixteenth inch, the percentage of precious metal is almost infinitesimal. To give you a comparison you can easily understand, it would take twenty-five layers (at one-ten-thousandth inch) of plated precious metal to make the thickness of an average sheet of writing paper. The base metal (at a sixteenth inch) would be 625 times thicker than the precious metal surface. To reclaim the precious metal, the plated item must be smelted. The resulting alloy from an item with all surfaces plated (if the plating is pure gold or silver) would have a possible fineness of between .0032 and .0048 (averaging .004, depending on the base metal). That would be about 996 parts out of 1000 of non-precious base metal. Of course, reclamation of plated precious metal items may be no more difficult than mining three tons of ore for one ounce of gold. With precious metals maintaining high prices, plated items *are* reclaimed. American Chemical and Refining Company specializes in such work, as do several other refineries.

Gold filled items are more suitable for reclamation. Gold filled has at least one-twentieth of the total weight in karat gold. Gold plate, gold overlay, and rolled gold plate usually have at least one-fortieth of the total weight in karat gold. This makes a far greater percentage of gold in the refined metal than with the electroplated items.

Still, it is very likely that a gatherer who is buying plated items is more interested in obtaining art pieces at scrap prices.

Regardless of what type of gold-covered item you might have, before even considering trading it for scrap, have it evaluated for its possible artistic value. You will often find the workmanship more valuable than the gold content of the finish. The same can be said for silver-plated items.

17

Gold Jewelry

The important point to remember about traditional gold jewelry is that it isn't an *investment* in gold. It is an investment in art. Traditional gold jewelry averages between one-sixth to one-third of the total cost in actual gold value. The other two-thirds to five-sixths of the cost pays for artistic work, taxes, retailer's overhead, and other related costs. The exception to this rule is quasi-investment jewelry. Depending on the premium you pay for it, such jewelry could be classified as an investment in gold. It still would command a higher premium than bullion or most bullion coins, but if you

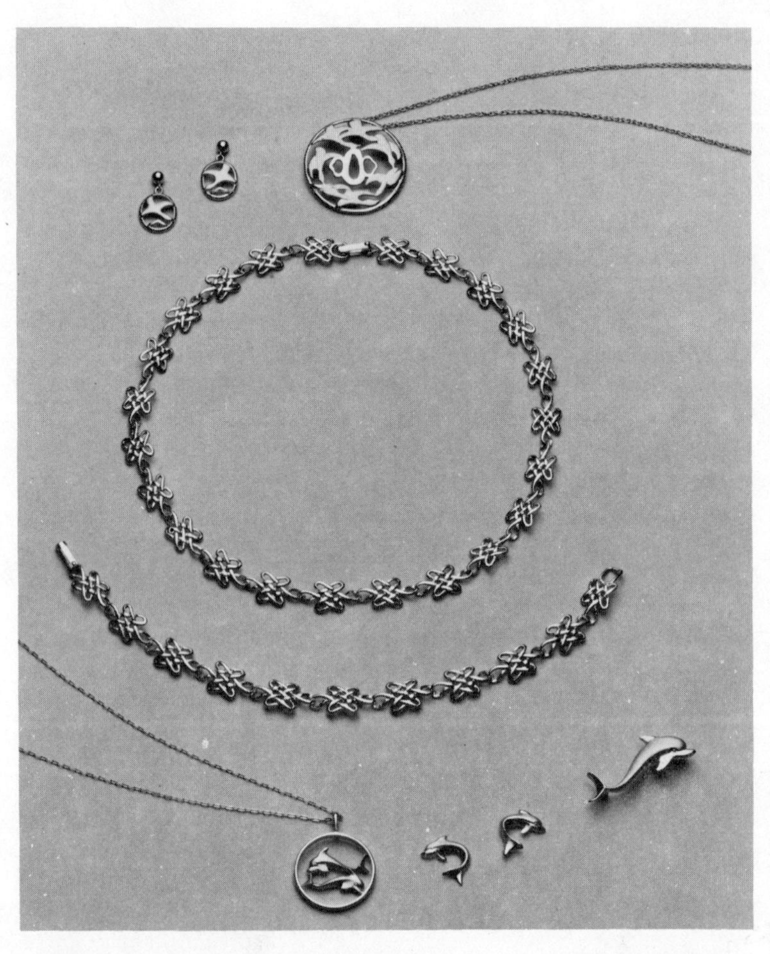

Examples of gold jewelry by Gorham Textron. (Photograph courtesy of Gorham Textron.)

wanted it for the the ownership of high-karat gold jewelry as opposed to an investment you couldn't wear, it would be worthwhile.

Traditionally, American jewelry is 14K gold. The European consumer is used to 18K gold, which has a richer look, and the preference for 18K is catching on in America.

In the Near and Far East, the quasi-investment jewelry has been popular for some time. It is only lately that American firms have started producing it. At least with this type of jewelry, you can realize a better return on your purchase if you are forced to sell it in hard times. Regular 14K, 18K, and other traditional kinds of jewelry would realize a considerable loss over the original purchase price. Any time jewelry is sold to a purchaser interested only in the gold content, only gold-weight value is paid, less the assaying-and-refining fee. You are likely to find, say, a bracelet for which you paid $450 bringing only $75, even if the spot price is the same at the time you bought and the time you sold the item. With quasi-investment jewelry, a piece purchased at $450 and sold later at the same spot price might return $325. There are several factors that could bear on either of these figures, so they are given as an example, not as a firm price.

Another important fact that you should remember about jewelry is the United States government-allowed discrepancies in karat markings. This is covered in the information above on karat gold.

18

Vermeil

Jewelers follow the French origin of the word in pronouncing it vair-may, *but your dictionary gives the pronunciation as* vur-mil, *similar to the word for the color vermilion, to which it is related.*

The word or product designation possibly most needing regulation is *vermeil.* If the government intends to clarify any misleading word applied to precious-metals products, this should be the one. After the government's failure with *solid gold,* however, I would hope any effort to clarify vermeil would be more successful. "Sometimes applied to precious metals . . ." is appropriately worded. Contrary to the opinion of many people, vermeil is not necessarily sterling silver covered or plated with (at least) karat gold. Vermeil can also be used to advertise a product made of stainless steel plated with copper. Or copper plated with 500-fine silver. Or brass plated with copper. Or—you name it. No standard exists.

Examples of vermeil jewelry by Gorham Textron. (Photograph courtesy of Gorham Textron.)

Some worthwhile suggestions have been made, for example, defining vermeil as a base of solid sterling silver with an electrolytically or mechanically bonded finish of karat gold at least 0.000012 or 0.000015 inch thick. This standard might be ideal. While there are many persons and firms in the precious-metals business trying to get some sort of regulation enacted, there are others who are happy with the prevailing situation. Proposals have been made for adequate regulation, and could be enacted within a reasonable time.

An example of a manufacturer applying the word *vermeil* with integrity is the "Golden Scroll" collection offered by Gorham Textron. Their jewelry, sterling silver electroplated with 22K gold, offers the look of gold without its high cost. An example of an item in the collection appears in the photograph.

A vermeil presentation by another well-known firm offers a product as gold vermeil—24K-gold electroplate. The base metal on this presentation is not sterling silver, however, but brass. The fact that brass is the base metal is not included in the advertisement. It required a phone call to the manufacturer to obtain this information.

The difference between these two examples points up the need for some standardization of the use of the word *vermeil.*

19

Miniature Replicas of Gold (and Silver) Coins

Replica gold and silver coins have been an increasingly advertised type of precious-metal product. For example, the Krugerrand replica in both 14K and 22K gold has been presented, along with 14K-gold Maple Leafs and 14K-gold United States $20 pieces. There may be others available or due to come out soon. While the advertisements for these coins may not be laid out intentionally to mislead the public, the average person would end up confused. The words *Solid Gold, 14-Karat Solid Gold, Solid 22-Karat Gold,* along with pictures of the actual (genuine) coins that read *Fyngoud 1 oz. Fine Gold, Fine Gold 1 oz. or Pur, Twenty Dollars,* leave a reader puzzled. The advertising copy goes on to describe the value of the genuine coin, and notes that you have to act within a specified time period to take advantage of the special price, owing to the constant fluctuation of the price of gold.

But just how critical is the price of gold to the production of these miniatures?

Obtaining the cooperation of the manufacturers of the miniatures was not easy, in fact, it was impossible. When telephoned, one miniature-coin manufacturer said that they couldn't provide information over the phone, please write a letter. The letter is still unanswered. About all that could be learned was gleaned from the answering services used to take the toll-free telephone inquiries. Both the Maple Leaf and Krugerrand miniatures were said to have a weight of .323 g. That was the extent of the information the answering services had that was not included in the newspaper ads.

With the karat designation and the .323 g. weight, along with the price being charged for the miniatures, the actual gold value can be determined. Since it is considered jewelry, rather than a coin, a miniature can be less than the karat percentage claimed by the manufacturer. The miniature-coin manufacturers would not tell me, when I phoned, if the miniatures were a full 14K or 22K, and did not reply to my letter requesting that information. I assume, therefore, that they didn't want the public to know. It is highly probable the miniatures are 13½K and 21½K, but for the sake of showing the greatest gold content possible, the following figures are based on plumb gold (gold labeled 14K that is, in fact, 14K). We know that the weight of the miniatures is 0.323 g. and that equals 0.0103845 troy oz.; and that 14K gold equals a fineness of 583.333, that is, there is 58.33 percent pure gold in the miniatures.

We also know that the genuine Maple Leaf weighs 31.1033 g. (1 full oz.) and is 999 fine (24K gold). Therefore, dividing the weight of the genuine Maple Leaf by the miniature (1 oz. ÷ 0.010384 oz. = 96.3020) shows that the miniature is less than one ninety-sixth of the genuine Maple Leaf. However, this accounts only for the total weight of the two coins. The genuine coin is pure gold (24K), while the miniature is only 58.33 percent pure gold (14K). The fine-gold content is 0.1884 g.: 323 g. × 0.5833 = 0.1884 g. Each miniature coin has 0.1884 g. pure gold. Again, if we divide the weight of the Maple Leaf in grams by the weight of the miniature in grams (31.1033 g. ÷ 0.1884 g. = 165.0918), we see that the pure-gold content of the miniature is $\frac{1}{165}$ of the genuine Maple Leaf. In other words, to equal one ounce of pure gold, you must purchase 165 miniatures; since each miniature costs $15, you would have to pay $2475—$2475 for an ounce of gold! Do you think it really matters if you send in your order in the specified time to "protect" the transaction from the constant fluctuation of the gold price? Even if the price of gold soared to over $2000 per ounce, the manufacturers would still be making a considerable profit.

This is not to say that $15 for a miniature Maple Leaf is an unreasonable price, however. Many reputable dealers offer the coins along with other novelties. These merchants allow the coins to go on their own merits. If you want to buy one at the set price, fine; there's no hypersalesmanship.

Possibly a better bargain is the miniature replica of the United States silver dollar; better, that is, if it wasn't hyped up with the inevitable come on—"only $1 if you act." Again the word *solid* is used, now in relation to sterling silver (92.5 percent pure silver). To take advantage of the one-dollar price, you have to buy ten replicas. We won't worry about the actual silver weight of a genuine silver dollar, since the value of the miniature must be balanced against a troy ounce of pure silver.

The sterling-silver dollar miniatures also weigh 0.323 g., or 0.010384 troy oz. If we divide one ounce by the weight of the replica (1 oz. ÷ 0.010384 = 96.3020), we see that it would take 96.3 miniatures to total one full ounce. Sterling silver, however, is 92.5 percent pure silver. If we multiply that percentage by the weight of the replica in grams (0.323 g. × 0.925 = 0.298775 g.), we see that each replica has 0.2988 g. pure silver. Dividing 31.104 g. (grams per ounce) by 0.2988 g. gives us the number of replica coins it takes to make one ounce of pure silver, 104.096. At one dollar for each miniature, the price of silver would have to zoom to twice the price of 1980's now-fabled high for the buyer to get his money's worth.

20

Numismatic Value

Most coins are minted for circulation as legal tender. As time goes on, however, many coins become collectors' items, gaining in worth far beyond their face value or precious metal content. Since man has been minting coins since antiquity, the coins collected by numismatists can vary greatly in age. They can vary greatly in condition too: Sometimes coins were deliberately withheld from circulation to increase their value over used and worn coins of the same issue. The value a coin acquires above its denomination and precious metal content is called its numismatic value, or its premium—this is an added value given to the coin by collectors because of its unusualness or desirability. (Every gold coin has some premium value, even those minted today—but rare coins sought after by numismatists can have great premium value far in excess of intrinsic worth.)

Basically the factors most influential in setting the numismatic premium for coin collectors are: the number of coins in the same issue (which usually, but not always, determines the coin's rarity); the condition of the coin, which is expressed in terms ranging from "about good" (AG-3) to "perfect uncirculated" (MS-70) and "proof"; and age or date of issue. Minting errors can also add to the value of any coin, even one minted today. There are coins issued by the Roman emperors that have a lower premium value than unusual coins minted in this decade. Because of this great variation of values, and the complexities of evaluating collectors' coins, it is of the utmost importance that you consult reliable coin dealers. A properly considered coin investment almost always realizes a continual climb in value, but if an investor approaches the matter with no regard to either advice from a reliable dealer and/or at least studying some of the many authoritative books, periodicals, magazines, and other publications available, it will be pure luck if the investment turns out favorably. I have found the book written by Q. David Bowers, *High Profits from Rare Coin Investment* (Los Angeles: Bowers and Reddy Galleries, 1977), to be exceptionally good. Bowers has written other books and articles, and I am sure that his work would be enlightening to a potential numismatist.

If you are buying a numismatic gold coin, the intrinsic gold content is almost incidental, and you may pay a large premium based on the coin's rarity, condition, demand, etc., rather than the dealer's smaller premium for handling and related charges in the trading of the coin. The following chart, for example, lists United States gold coins, which traditionally have *high* numismatic value that places them out of the bullion category. These United States gold coins *are* a good investment *if* bought within the prevailing price range for those coins. You have to be sure you aren't paying an inflated numismatic price, however. The likelihood of paying an inflated price is diminished if you deal with firms of high integrity. For a comparison of the value of bullion coins with the United States gold coins, take a United States 20-dollar gold piece with a fine-gold content of .9675 oz., almost a full ounce of gold. The South African Krugerrand (often miscalled Krugerr*land*) has a fine-gold content of one ounce. The Krugerrand price fluctuates with the spot gold price. If gold is selling at $700 an ounce, that is the value of the Krugerrand. The United States 20-dollar gold piece, however, can cost over $5000 (depending on the specific issue of coin). If you are interested in coins numismatically, or as an investment to tie up your capital in a very small item that is easily stored (but requires careful security and insurance), then go with a numismatic coin.

The chart on United States gold coins is carried to decimal positions past the usual four places. Actually, United States gold

United States Gold Coins—All Denominations

COIN	DATES OF ISSUE	ACTUAL COIN WEIGHT	FINENESS	WEIGHT OF GOLD CONTENT
One Dollar Piece	1849–1859	0.05375 oz.	900	0.048375 oz.
Three Dollar Piece	1854–1899	0.16125 oz.	916⅔	0.1478124 oz.
$2.50 Quarter Eagle	1796–1834	0.140625 oz.	916⅔	0.1289062 oz.
$2.50 Quarter Eagle	1834–1839	0.134375 oz.	899.225	0.1208333 oz.
$2.50 Quarter Eagle	1840–1929	0.134375 oz.	900	0.1209375 oz.
$5.00 Half Eagle	1795–1834	0.28125 oz.	916⅔	0.2578124 oz.
$5.00 Half Eagle	1834–1838	0.26875 oz.	899.225	0.2416667 oz.
$5.00 Half Eagle	1837 only	0.26875 oz.	900	0.241875 oz.
$5.00 Half Eagle	1839–1929	0.26875 oz.	900	0.241875 oz.
$10.00 Eagle	1795–1804	0.5625 oz.	916⅔	0.5156249 oz.
$10.00 Eagle	1838–1866	0.5375 oz.	899.225	0.4833334 oz.
$10.00 Eagle	1837 only	0.5375 oz.	900	0.48375 oz.
$10.00 Eagle	1866–1933	0.5375 oz.	900	0.48375 oz.
$20.00 Double Eagle	1849–1933	1.075 oz.	900	0.9675 oz.

United States gold coins have a traditionally high numismatic value. Before trading any gold coin as a consideration of the surge in precious-metals dealings, first check its numismatic value.

coin has an assortment of weights and finenesses. Many publications round off the fine designations to 900 fine, but there are actually three fineness figures to go with the assorted weights in different denominations.

If a coin with high numismatic premium value becomes damaged so badly that it no longer has a numismatic value, the gold in the coin must be treated as scrap, and the value of the coin goes down to its precious metal value, just as the value of modern bullion coinage is determined. If the damaged coin is sold for scrap, its value varies with the spot price of gold, and deductions should be made for refining charges and losses. In other words, its numismatic or premium value is gone and its intrinsic gold value prevails.

To give an example of how to calculate its market value, we'll use the Holland 10 Guilders, even though this is not a numismatic coin, but a legal tender gold coin for which we know its fineness and gold content. The method of calculation is always the same, once you know the fineness and weight of the coin. In the case of a damaged numismatic coin, you have to consult an appropriate coin collector's catalog that lists the fineness.

The chart on page 60 gives the fineness of the Holland 10 Guilders as 900, with a fine-gold weight of 0.1947 oz. Assuming the spot gold price is \$700, the full value of the fine gold content is \$136.29 (\$700 × 0.1947).

If we deduct 2 percent for refinery losses, and \$0.60 per oz. for refining charges, the market value of the Holland 10 Guilders becomes: \$133.44 = \$136.29 − \$2.73 (loss) − \$0.12 (refining).

Therefore, \$133.44 is the most this coin can fetch for its scrap gold value at the above spot gold price.

21

Contemporary Gold Bullion Coins

Although the dictionary definition of *bullion* restricts usage of the word to bars and ingots, today the precious-metals industry uses the word for *any* gold (or silver) form that is traded solely for its intrinsic precious-metal value. The word is even applied to 22K or 24K jewelry produced for investment purposes. If a coin is traded for its face value or its numismatic worth, it cannot be considered a bullion coin.

Bullion coins are minted for their gold content and stored in uncirculated condition. The value of a contemporary bullion coin is set almost entirely by the latest spot gold price—with a slight surcharge premium added by the dealer (see the chart on page 60). That spot price can, and often does, change several times daily. Many coin dealers have computers tied in with all the major gold markets. At any time during the trading day they can punch in London, Zurich, or any center they desire, getting the spot price current at that moment. That price is the one you should pay, plus the specific premium for the coin of your choice.

You must realize that, like any gold investment, a bullion coin will drop in price if the spot price of gold goes down. Conversely, if the spot price goes up, the value of the bullion coin will rise accordingly.

Bullion coins are not as negotiable as regular money, but they are relatively easily traded in many countries—if your coins have been maintained in their original purchase condition. A damaged bullion coin loses value well below the spot gold price. Damaged gold coins are examined with great care, and if the dealer determines that the damage (nick, scratch, etc.) is too extensive, you will be unable to trade the coin on the spot. The coin could then be treated as scrap gold, requiring a melt-and-assay process before you could receive payment. You would also have to pay the costs for that operation. Bullion coins traded by reliable dealers are protected in envelopes of clear plastic. Leave your coin in that wrapper. When you trade a bullion coin, you are most often paid the spot gold price only, or the spot price plus a portion of the premium.

The advantage of the size of a bullion coin, ranging from a fraction of an ounce to a full ounce of pure-gold content, makes the bullion coin accessible to the person with only a small amount of initial capital. Gold bullion bars are a better investment because there is either no premium cost involved, or the premium cost is lower than the premium for gold coins. The weight of a bar requires a higher intrinsic investment, however, and many small investors cannot afford purchasing a one-kilogram bar. Of course, many dealers now carry

Commonly Traded Contemporary Gold Coins and Their Premiums

COUNTRY AND COIN	TOTAL COIN WEIGHT	FINE-NESS	WEIGHT OF FINE GOLD	AVERAGE PREMIUM
England—One Pound (original)	0.2568 oz.	916⅔	0.2354 oz.	16–18%
France—20 Francs (original)	0.2074 oz.	900	0.1867 oz.	24–25%
Holland—10 Guilders (original)	0.2163 oz.	900	0.1947 oz.	11–12%
Belgium—20 Francs (original)	0.2074 oz.	900	0.1867 oz.	11–11½%
Italy—20 Lire (original)	0.2074 oz.	900	0.1867 oz.	15–16%
Switzerland—20 Francs (original)	0.2074 oz.	900	0.1867 oz.	15–16%
Russia—10 Rubles (restrike with new dates)	6.2767 oz.	900	0.2489 oz.	3–3½%
Austria—100 Koronas (restrike)	1.0891 oz.	900	0.9802 oz.	2–2½%
20 Koronas (restrike)	0.2178 oz.	900	0.1960 oz.	7–7½%
10 Koronas (restrike)	0.1089 oz.	900	0.0980 oz.	8–9%
4 Ducats (restrike)	0.4489 oz.	986	0.4430 oz.	4–4½%
1 Ducat (restrike)	0.1122 oz.	986	0.1107 oz.	6½–7%

Hungary—100 Koronas (restrike)	1.0891 oz.	900	0.9802 oz.	2–2½%
South Africa—Krugerrand (original)	1.0909 oz.	916⅔	1.0000 (full oz.)	4½–5%
2 Rands (original)	0.2568 oz.	916⅔	0.2354 oz.	7–7½%
1 Rand (original)	0.1284 oz.	916⅔	0.1177 oz.	11–12%
Canada—Maple Leaf (original)	1.0000 (full oz.)	999	1.0000 (full oz).	3–3½%
Chile—100 Pesos (original)	0.6539 oz.	900	0.5885 oz.	2–2½%
Colombia—5 Pesos (original)	0.2568 oz.	916⅔	0.2354 oz.	1–1½%
Peru—1 Libra (original)	0.2568 oz.	916⅔	0.2354 oz.	3–3½%
Mexico—50 Pesos (restrike)	1.3396 oz.	900	1.2057 oz.	4–4½%
20 Pesos (restrike)	0.5358 oz.	900	0.4823 oz.	6–6½%
10 Pesos (restrike)	0.2679 oz.	900	0.2411 oz.	7–7½%
5 Pesos (restrike)	0.1339 oz.	900	0.1205 oz.	9–10%
2½ Pesos (restrike)	0.0670 oz.	900	0.0603 oz.	11–13%
2 Pesos (restrike)	0.0536 oz.	900	0.0482 oz.	14–16%

(This data is courtesy of Manfra, Tordella & Brookes, Inc.)

the Swiss Credit Bank gold bars in weights ranging from as little as five grams (0.1607 oz.) to one kilogram (32.15 oz.), making smaller investments in bar-form bullion possible. The graduated size of this bullion allows a purchaser to find a bar to "fit his pocketbook."

The attractiveness of gold coins for investors, other than the actual beauty of the coins, is in the convenient size. Many people wish to actually take possession of their gold holdings, and coins are well suited to that. They can be picked up, carried, and stored with ease.

If you plan to invest in precious metal beyond the simple purchase of a coin or two, it is strongly recommended that you consult an expert in the field in which you are interested. Such authorities can advise you on more than just the hard gold—for example, on tax law which has a direct bearing on potential profit. You must realize that gold, unlike stocks, bonds, and other types of investments, does not have an interest-creating system. If you buy at $620 an ounce and a year later the price is $518 an ounce, your investment has produced no profits, in fact, you've lost money. Even if the spot price is up to $678 an ounce, you might still have lost money. Insurance, premium charges, storage costs, and other factors might have eaten away at the spot-price profit.

Any purchase of gold coins should be made from a reliable dealer. Counterfeit coins are not uncommon, and your only protection is the integrity of the dealer. It takes an expert to detect a well-made counterfeit, and what appears to be an authentic bullion coin (counterfeits abound in numismatic coins also) with a fineness of 900 might in fact be an exact copy with a fineness of only 500.

The following companies can be a starting point for your interest in gold bullion and/or coin investment:

Deak-Perera
29 Broadway
New York, N.Y. 10006
Phone: (212) 480-0200 or (212) 757-6915

This firm deals in all the popular bullion-type gold coins, 24K-gold bullion jewelry, bullion bars, precious-metals certificates, international currency, travelers checks, and several other financial services. There are several offices in the New York City area, as well as office locations in California (six locations); Connecticut (two locations); District of Columbia; Florida; Guam (two locations); Hawaii (three locations); Illinois; New York (six locations); Puerto Rico; Saipan; Canada (four locations); Austria; Switzerland (two locations); England; Hong Kong; Macao; Alaska; Texas; Israel; West Germany.

Manfra, Tordella & Brookes, Inc.
59 W. 49th St.
New York, N.Y. 10020
Phone: (212) 974-3400 or (212) 775-1440

This firm deals in all the popular bullion-type gold coins, a wide range of weights in bullion bars, silver coins and bars, silver and gold numismatic coins, foreign currencies, and other precious-metals items. There is an office in lower Manhattan at 151 World Trade Center Concourse.

Either of these two firms will deal by phone, mail, or in person. If you telephone, keep your conversation short and to the point. They receive many calls a day and their traders are kept very busy. It may be necessary, or beneficial to you in some cases, to deal with precious-metals firms through your bank.

Regardless of where you live, you might find it just as convenient to deal with a larger firm in a major metropolitan area—such as the two listed above—as with regional firms. Still, there are many highly qualified precious-metals dealers found in even the smallest cities. The main point for you to remember is to deal only with well-established *and* reputable dealers. Your personal banker could be a valuable consultant in these matters.

Either of the firms mentioned will send their literature upon request; Deak-Perera can provide a directory of their many office locations.

The preceding chart lists all the popular gold bullion coins, along with the average premiums charged (over the spot gold price). The premiums charged may and will vary from the percentages listed, going up or down with several bearing factors, such as supply and demand. Occasionally, a dealer may be out of specific coins, but there is usually a wide selection from which to choose. If you are interested only in the bullion coins as an investment, you should choose those with the lowest premium charge. Usually the smaller coins, such as the Holland 10 Guilder (with a total coin weight of .2163 oz. and fine gold weight of .1947 oz.) have a higher premium cost. The Mexican 10 Peso, with a fine-gold content of .25 oz., commands an average 7½ percent premium, while the Mexican 50 Peso, with a fine-gold content of 1.2 oz. requires an average premium of 4½ percent.

Almost all United States gold coins have a numismatic value that places them out of the bullion-coin category (see page 57).

In addition to the bullion coins listed on the preceding chart, the following privately minted coins are traded as bullion gold coins:

More Bullion Gold Coins

Engelhard Prospector	Total weight:	1 full oz.
	Fineness:	9995
	Gold Weight:	1 oz.
	Average prem.:	6 to 7 percent
FRANKLIN MINT GOLD PIECES:		
Franklin Mint 1 oz.	Total weight:	1.001 oz.
	Fineness:	999
	Gold Weight:	1.000 oz.
	Average prem.:	3 to 3½ percent
Franklin Mint 0.5 oz.	Total weight:	0.5005 oz.
	Fineness:	999
	Total weight:	0.5 oz.
	Average prem.:	4 to 4½ percent
Franklin Mint 0.25 oz.	Total weight:	0.25025 oz.
	Fineness:	999
	Gold Weight:	0.25 oz.
	Average Prem.:	8 to 8½ percent

(This data courtesy of Manfra, Tordella & Brookes, Inc.)

The premium is higher on the Engelhard Prospector coin because it is no longer being minted by Engelhard, and thus the supply is less than coins still being minted, such as the Krugerrand.

The following charts list gold coins issued as legal tender worldwide in 1979 (the latest year complete figures are available). Government interest in gold coinage is increasing each year. In 1977, 46 countries issued 87 different coins. In 1978, that figure climbed to 49 countries issuing 128 different coins. The following data on the year 1979 shows 80 nations issuing a total of 230 different coins.

Worldwide Gold Coinage in 1979†

COUNTRY	COIN	COIN WEIGHT	FINE-NESS	WEIGHT OF FINE GOLD	NUMBER OF COINS ISSUED
Afghanistan	10,000 Afghanis	1.0750 oz.	900	0.9675 oz.	168
Andorra	1 Sovereign	0.2572 oz.	918	0.2361 oz.	500
Bahamas	* 2,500 Dollars	13.0938 oz.	916.6	12.0017 oz.	5
* *(2 issues—same design, different dates)*	* 2,500 Dollars	13.0938 oz.	916.6	12.0017 oz.	47
	250 Dollars	0.3401 oz.	900	0.3061 oz.	1,585
** *(2 issues—2 designs)*	** 100 Dollars	0.4375 oz.	916.7	0.4011 oz.	2,822
	** 100 Dollars	0.4375 oz.	916.7	0.4011 oz.	1,566
Barbados	200 Dollars	0.3247 oz.	900	0.2922 oz.	500
	200 Dollars	0.2611 oz.	900	0.2350 oz.	500
	100 Dollars	0.1624 oz.	900	0.1461 oz.	500
	100 Dollars	0.1305 oz.	900	0.1175 oz.	500
Belize	* 100 Dollars	0.1997 oz.	500	0.0998 oz.	4,865
* *(2 issues—2 designs)*	* 100 Dollars	0.1997 oz.	500	0.0998 oz.	6,440
Benin	10,000 Francs	1.1429 oz.	900	1.0286 oz.	30
	5,000 Francs	0.5713 oz.	900	0.5142 oz.	40
	2,500 Francs	0.2855 oz.	900	0.2569 oz.	30
Botswana	150 Pulas	1.0750 oz.	900	0.9675 oz.	215
	150 Pulas	0.5138 oz.	916.6	0.4709 oz.	473
British Virgin Islands	100 Dollars	0.2283 oz.	900	0.2054 oz.	3,216
Brunei	1,000 Dollars	1.6075 oz.	916.6	1.4734 oz.	273

† (Courtesy of the Gold Institute, and adapted from its publication, *Modern Gold Coinage.*)

Worldwide Gold Coinage in 1979—Continued

COUNTRY	COIN	COIN WEIGHT	FINE-NESS	WEIGHT OF FINE GOLD	NUMBER OF COINS ISSUED
Canada	100 Dollars	0.5454 oz.	916.6	0.5000 oz.	250,000
	50 Dollars Maple Leaf	1.0000 oz.	999.9	1.0000 oz.	1,000,000
Cayman Islands	100 Dollars	0.7292 oz.	500	0.3646 oz.	1,198
* *(6 issues—6 designs)*	* 50 Dollars	0.3646 oz.	500	0.1823 oz.	812
	* 50 Dollars	0.3646 oz.	500	0.1823 oz.	806
	* 50 Dollars	0.3646 oz.	500	0.1823 oz.	786
	* 50 Dollars	0.3646 oz.	500	0.1823 oz.	578
	* 50 Dollars	0.3646 oz.	500	0.1823 oz.	572
	* 50 Dollars	0.3646 oz.	500	0.1823 oz.	547
	100 Dollars	0.7292 oz.	500	0.3646 oz.	1,198
Chile	500 Pesos	3.2697 oz.	900	2.9427 oz.	200
	100 Pesos	0.6539 oz.	900	0.5885 oz.	150,000
	50 Pesos	0.3270 oz.	900	0.2943 oz.	850
	20 Pesos	0.1309 oz.	900	0.1178 oz.	30,000
	Onza Troy	1.0000 oz.	999.9	1.0000 oz.	1,580
China—People's Republic	450 Renminbi	0.5520 oz.	900	0.4968 oz.	50,000
* *(4 issues—4 designs)*	* 400 Yuan	0.5433 oz.	916.67	0.4981 oz.	70,000
	* 400 Yuan	0.5433 oz.	916.67	0.4981 oz.	70,000
	* 400 Yuan	0.5433 oz.	916.67	0.4981 oz.	70,000
	* 400 Yuan	0.5433 oz.	916.67	0.4981 oz.	70,000
Cook Islands	200 Dollars	0.5337 oz.	900	0.4803 oz.	2,235
	100 Dollars	0.3138 oz.	900	0.2824 oz.	3,792

Costa Rica	1,500 Colones	1.0750 oz.	900	0.9675 oz.	2
Cyprus	50 Pounds	0.5138 oz.	916.6	0.4709 oz.	18,100
Czechoslovakia	1 Duak	0.1122 oz.	986	0.1107 oz.	10,000
Dominica	* 300 Dollars	0.6173 oz.	900	0.5556 oz.	800
* *(3 issues—2 designs, 2 Mints)*	* 300 Dollars	0.6173 oz.	900	0.5556 oz.	800
** *(2 issues—1 design, 2 Mints)*	* 300 Dollars	0.6173 oz.	900	0.5556 oz.	100
	** 150 Dollars	0.3086 oz.	900	0.2778 oz.	450
	** 150 Dollars	0.3086 oz.	900	0.2778 oz.	134
Dominican Republic	250 Pesos	0.9999 oz.	900	0.8999 oz.	4,000
	100 Pesos	0.3858 oz.	900	0.3472 oz.	4,000
El Salvador	200 Colones	0.7587 oz.	900	0.6829 oz.	20
	100 Colones	0.3794 oz.	900	0.3414 oz.	50
	50 Colones	0.1897 oz.	900	0.1707 oz.	30
	25 Colones	0.0948 oz.	900	0.0854 oz.	2,000
Ethiopia	600 Birr	1.0750 oz.	900	0.9675 oz.	643
Equatorial Guinea	* 10,000 Ekuele	0.4475 oz.	916.9	0.4103 oz.	31
* *(2 issues—2 designs)*	* 10,000 Ekuele	0.4475 oz.	916.9	0.4103 oz.	68
	5,000 Ekuele	0.2238 oz.	916.9	0.2052 oz.	68
	1,000 Pesetas	0.4533 oz.	900	0.4080 oz.	110
	750 Pesetas	0.3398 oz.	900	0.3058 oz.	200
	500 Pesetas	0.2267 oz.	900	0.2040 oz.	100
	250 Pesetas	0.1132 oz.	900	0.1019 oz.	600
Falkland Islands	150 Pounds	1.0750 oz.	900	0.9675 oz.	674
Fiji	250 Dollars	1.0750 oz.	900	0.9675 oz.	534
France	50 Francs	3.2793 oz.	920	3.0170 oz.	417
	10 Francs	1.2506 oz.	920	1.1506 oz.	312
	5 Francs	1.2506 oz.	920	1.1506 oz.	312

Worldwide Gold Coinage in 1979—Continued

COUNTRY	COIN	COIN WEIGHT	FINE-NESS	WEIGHT OF FINE GOLD	NUMBER OF COINS ISSUED
	2 Francs	0.9934 oz.	920	0.9140 oz.	612
	1 Franc	0.7555 oz.	920	0.6591 oz.	312
	½ Franc	0.5948 oz.	920	0.5472 oz.	312
	20 Centimes	0.5626 oz.	920	0.5176 oz.	312
	10 Centimes	0.4180 oz.	920	0.3845 oz.	312
	5 Centimes	0.2797 oz.	920	0.2573 oz.	312
	1 Centime	0.2411 oz.	920	0.2218 oz.	312
French Polynesia	100 Francs	1.3535 oz.	920	1.2452 oz.	254
	50 Francs	2.0319 oz.	920	1.8693 oz.	184
	20 Francs	1.353 oz.	920	1.2452 oz.	184
	10 Francs	0.8134 oz.	920	0.7483 oz.	184
	5 Francs	1.5754 oz.	920	1.4493 oz.	184
	2 Francs	0.9902 oz.	920	0.9110 oz.	184
	1 Franc	0.5851 oz.	920	0.5383 oz.	184
Gambia	500 Dalasi	1.0750 oz.	900	0.9675 oz.	144
Gibraltar	100 Pounds	1.0000 oz.	916.6	0.9166 oz.	100
	50 Pounds	0.5000 oz.	916.6	0.4583 oz.	100
	25 Pounds	0.2500 oz.	916.6	0.2291 oz.	1,800
Great Britain	1 Souvereign	0.2568 oz.	916.6	0.2354 oz.	7,447,000
Greece	10,000 Drachmas	0.6430 oz.	900	0.5787 oz.	10,000
Guinea	5,000 Francs	0.6430 oz.	800	0.5787 oz.	750
Haiti	500 Gourdes	0.2733 oz.	900	0.2459 oz.	350
	500 Gourdes	0.2733 oz.	900	0.2459 oz.	350
	100 Gourdes	0.6350 oz.	900	0.5715 oz.	180

Hong Kong	* 1,000 Dollars	0.5136 oz.	916.6	0.4708 oz.	21,970
* *(5 issues—5 designs)*	* 1,000 Dollars	0.5136 oz.	916.6	0.4708 oz.	14,788
	* 1,000 Dollars	0.5136 oz.	916.6	0.4708 oz.	1,318
	* 1,000 Dollars	0.5136 oz.	916.6	0.4708 oz.	1,915
	* 1,000 Dollars	0.5136 oz.	916.6	0.4708 oz.	1,361
Iraq	100 Dinars	0.8359 oz.	916.6	0.7662 oz.	10,000
	50 Dinars	0.4405 oz.	916.6	0.4037 oz.	10,000
Isle of Man	5 Pounds	1.2799 oz.	916.7	1.1733 oz.	2,000
	2 Pounds	0.5118 oz.	916.7	0.4692 oz.	32,000
	1 Pound	0.2559 oz.	916.7	0.2346 oz.	60,000
	½ Pound	0.1280 oz.	916.7	0.1173 oz.	38,000
	1 Crown	1.3825 oz.	916.7	1.2673 oz.	300
Jamaica	* 250 Dollars	1.3889 oz.	900	1.2500 oz.	276
* *(2 issues—2 designs)*	* 250 Dollars	1.3889 oz.	900	1.2500 oz.	1,650
** *(2 issues—2 designs)*	** 100 Dollars	0.3646 oz.	900	0.3281 oz.	803
	** 100 Dollars	0.3646 oz.	900	0.3281 oz.	3,000
Jordan	50 Dinars	1.0772 oz.	900	0.9675 oz.	10
Kenya	3,000 Shillings	1.2860 oz.	916.6	1.1787 oz.	2,000
Kiribati	150 Dollars	0.5138 oz.	916.6	0.4709 oz.	2,379
Lesotho	250 Maloti	1.0909 oz.	916.7	1.0000 oz.	500
	250 Maloti	1.0909 oz.	916.6	0.9999 oz.	500
Liberia	100 Dollars	0.3514 oz.	900	0.3163 oz.	1,656
Macao	500 Patacas	0.2559 oz.	916	0.2344 oz.	5,500
Malawi	250 Kwacha	1.0750 oz.	900	0.9675 oz.	621
Malaysia	500 Ringgits	1.0750 oz.	900	0.9675 oz.	34
Malta	100 Pounds	1.0275 oz.	916	0.9412 oz.	4,750
	50 Pounds	0.5138 oz.	916	0.4706 oz.	4,750
	25 Pounds	0.2569 oz.	916	0.2353 oz.	4,750

Worldwide Gold Coinage in 1979—Continued

COUNTRY	COIN	COIN WEIGHT	FINE-NESS	WEIGHT OF FINE GOLD	NUMBER OF COINS ISSUED
Mauritius	1,000 Rupees	1.0750 oz.	900	0.9675 oz.	1
	100 Rupees	0.5138 oz.	916.6	0.4709 oz.	200
Mexico	50 Pesos	1.3396 oz.	900	1.2056 oz.	1,060,000
	20 Pesos	0.5358 oz.	900	0.4823 oz.	112,000
	10 Pesos	0.2679 oz.	900	0.2411 oz.	135,500
	5 Pesos	0.1340 oz.	900	0.1206 oz.	298,500
	2½ Pesos	0.0670 oz.	900	0.0603 oz.	604,000
	2 Pesos	0.0536 oz.	900	0.0482 oz.	611,250
Mongolia	750 Tukhrit	1.0750 oz.	900	0.9675 oz.	52
Morocco	500 Dirhams	0.5154 oz.	900	0.4639 oz.	3,300
* *(3 issues—3 designs)*	500 Dirhams	0.6221 oz.	900	0.5599 oz.	20
** *(2 issues—2 designs)*	250 Dirhams	0.2126 oz.	900	0.1913 oz.	906
	* 50 Dirhams	1.9335 oz.	900	1.7402 oz.	70
	** 50 Dirhams	2.9003 oz.	900	2.6102 oz.	13
	* 50 Dirhams	1.9335 oz.	900	1.7402 oz.	70
	** 50 Dirhams	2.9003 oz.	900	2.6102 oz.	30
	* 50 Dirhams	1.9335 oz.	900	1.7402 oz.	70
Nepal	Asarfi	0.3215 oz.	100	0.3215 oz.	52
	Asarfi	0.1608 oz.	100	0.1608 oz.	36
	Asarfi	1.0750 oz.	900	0.9675 oz.	13
Netherlands Antilles	50 Guilders	0.1080 oz.	900	0.0972 oz.	75,000
New Caledonia	100 Francs	1.3535 oz.	920	1.2452 oz.	254
	50 Francs	2.0319 oz.	920	1.8693 oz.	184

	20 Francs	1.3535 oz.	920	1.2452 oz.	184
	10 Francs	0.8134 oz.	920	0.7483 oz.	184
	5 Francs	1.5754 oz.	920	1.4493 oz.	184
	2 Francs	0.9902 oz.	920	0.9110 oz.	184
	1 Franc	0.5851 oz.	920	0.5383 oz.	184
New Hebrides	50 Francs	2.0319 oz.	920	1.8693 oz.	204
	20 Francs	1.3535 oz.	920	1.2452 oz.	204
	10 Francs	0.8134 oz.	920	0.7483 oz.	204
	5 Francs	1.5754 oz.	920	1.4493 oz.	204
	2 Francs	0.9902 oz.	920	0.9110 oz.	204
	1 Franc	0.5851 oz.	920	0.5383 oz.	204
Pakistan	3,000 Rupees	1.0750 oz.	900	0.9675 oz.	41
Panama	500 Balboas	1.3407 oz.	900	1.2066 oz.	1,787
	100 Balboas	0.2623 oz.	900	0.2361 oz.	5,079
Papua New Guinea	100 Kinas	0.3077 oz.	900	0.2769 oz.	3,880
Paraguay	10 Guaranies	0.2556 oz.	916.7	0.2343 oz.	20
	5 Guaranies	0.2775 oz.	916.7	0.2543 oz.	20
	1 Guarani	0.2058 oz.	916.7	0.1886 oz.	20
Peru	* 100,000 Soles	1.0912 oz.	916.6	1.0002 oz.	10,000
* *(3 issues—3 designs)*	* 100,000 Soles	1.0912 oz.	916.6	1.0002 oz.	10,000
** *(3 issues—3 Designs)*	* 100,000 Soles	1.0912 oz.	916.6	1.0002 oz.	10,000
	** 50,000 Soles	0.5456 oz.	916.6	0.5001 oz.	10,000
	** 50,000 Soles	0.5456 oz.	916.6	0.5001 oz.	10,000
	** 50,000 Soles	0.5456 oz.	916.6	0.5001 oz.	10,000
Poland	* 2,000 Zlotych	0.2588 oz.	900	0.2329 oz.	5,000
* *(3 issues—3 designs)*	* 2,000 Zlotych	0.2588 oz.	900	0.2329 oz.	5,000
	* 2,000 Zlotych	0.2588 oz.	900	0.2329 oz.	268

Worldwide Gold Coinage in 1979—Continued

COUNTRY	COIN	COIN WEIGHT	FINE-NESS	WEIGHT OF FINE GOLD	NUMBER OF COINS ISSUED
San Marino	5 Scudi	0.4823 oz.	917	0.4422 oz.	21,000
	2 Scudi	0.1929 oz.	917	0.1769 oz.	37,000
	1 Scudo	0.0965 oz.	917	0.0884 oz.	37,000
Sao Tome and Principe	* 2,500 Dobras	0.2083 oz.	900	0.1875 oz.	100
* *(5 issues—5 designs)*	* 2,500 Dobras	0.2083 oz.	900	0.1875 oz.	100
	* 2,500 Dobras	0.2083 oz.	900	0.1875 oz.	100
	* 2,500 Dobras	0.2083 oz.	900	0.1875 oz.	100
	* 2,500 Dobras	0.2083 oz.	900	0.1875 oz.	100
Seychelles	1,500 Rupees	1.0750 oz.	900	0.9675 oz.	47
Solomon Islands	100 Dollars	0.3012 oz.	900	0.2711 oz.	49
Somalia	1,500 Shillings	0.5138 oz.	916	0.4706 oz.	175
South Africa	2 Rand	0.2568 oz.	916.7	0.2355 oz.	32,000
	1 Rand	0.1284 oz.	916.7	0.1177 oz.	54,400
	Krugerrand	1.0909 oz.	916.7	1.0000 oz.	4,700,511
Sudan	100 Pounds	1.0750 oz.	900	0.9675 oz.	63
	50 Pounds	0.5626 oz.	916.9	0.5159 oz.	222
	25 Pounds	0.2652 oz.	916.3	0.2430 oz.	482
Swaziland	2 Emalangeni	0.9999 oz.	999	1.0000 oz.	1,250
	1 Lilangeni	0.4999 oz.	999	0.5000 oz.	1,250
Thailand	5,000 Baht	1.0750 oz.	900	0.9675 oz.	89
Tunisia	10 Dinars	0.6047 oz.	900	0.5442 oz.	2,000

Turks & Caicos Islands	100 Crowns	0.5787 oz.	500	0.2894 oz.	2,000
* *(10 issues—10 designs)*	100 Crowns	0.5792 oz.	500	0.2896 oz.	540
	* 50 Crowns	0.2896 oz.	500	0.1448 oz.	270
	* 50 Crowns	0.2896 oz.	500	0.1448 oz.	268
	* 50 Crowns	0.2896 oz.	500	0.1448 oz.	269
	* 50 Crowns	0.2896 oz.	500	0.1448 oz.	265
	* 50 Crowns	0.2896 oz.	500	0.1448 oz.	266
	* 50 Crowns	0.2896 oz.	500	0.1448 oz.	266
	* 50 Crowns	0.2896 oz.	500	0.1448 oz.	265
	* 50 Crowns	0.2896 oz.	500	0.1448 oz.	261
	* 50 Crowns	0.2896 oz.	500	0.1448 oz.	254
	* 50 Crowns	0.2896 oz.	500	0.1448 oz.	266
Turkey	* 1,000 Liras	0.5144 oz.	916.7	0.4716 oz.	650
* *(3 issues—3 designs)*	* 1,000 Liras	0.5144 oz.	916.7	0.4716 oz.	500
** *(4 issues—4 designs)*	** 500 Liras	0.2572 oz.	916.7	0.2358 oz.	650
	** 500 Liras	0.2572 oz.	916.7	0.2358 oz.	900
	** 500 Liras	0.2572 oz.	916.7	0.2358 oz.	900
	* 1,000 Liras	0.5144 oz.	916.7	0.4716 oz.	783
	** 500 Liras	0.2572 oz.	916.7	0.2358 oz.	783
Tuvalu	50 Dollars	0.5138 oz.	916.6	0.4709 oz.	200
Union of the Soviet Socialist Republics	100 Roubles	0.5556 oz.	900	0.5000 oz.	130,000
	10 Roubles	0.2766 oz.	900	0.2489 oz.	1,000,000
Uruguay	5 Centesimos	0.4035 oz.	900	0.3631 oz.	52
	2 Centesimos	0.2974 oz.	900	0.2676 oz.	52
	1 Centesimo	0.2013 oz.	900	0.1811 oz.	52
Venezuela	1,000 Bolivares	1.0750 oz.	900	0.9675 oz.	182
Western Samoa	100 Dollars	0.4019 oz.	916	0.3681 oz.	1,000

Worldwide Gold Coinage in 1979—Concluded

COUNTRY	COIN	COIN WEIGHT	FINE-NESS	WEIGHT OF FINE GOLD	NUMBER OF COINS ISSUED
Yemen Arab Republic	50 Riyals	1.5754 oz.	900	1.4178 oz.	103
* *(2 issues—2 designs)*	30 Riyals	0.9452 oz.	900	0.8507 oz.	100
	* 20 Riyals	0.6301 oz.	900	0.5671 oz.	100
	* 20 Riyals	0.6301 oz.	900	0.5671 oz.	100
	10 Riyals	0.3151 oz.	900	0.2836 oz.	200
	5 Riyals	0.1575 oz.	900	0.1418 oz.	200
Yugoslavia	5,000 Dinars	0.9484 oz.	900	0.8536 oz.	10,052
	2,500 Dinars	0.4726 oz.	900	0.4253 oz.	13,462
	2,000 Dinars	0.3794 oz.	900	0.3414 oz.	10,898
	1,500 Dinars	0.2829 oz.	900	0.2546 oz.	12,112
Zaire	100 Zaires	1.0750 oz.	900	0.9675 oz.	43
Zambia	250 Kwacha	1.0750 oz.	900	0.9675 oz.	640

Studying these charts, you will find some of the bullion coins, such as the Krugerrand and Mexican Peso. You will also find a large number of issues with very small total coinage. These coins are of special interest, since many are proof coins only. For more complete information on modern gold coins, write to the Gold Institute/L'Institut de l'Or, 1001 Connecticut Avenue, N.W., Washington, D.C., 20036, including ten dollars with your order for the book *Modern Gold Coinage.* This book is the most authoritative publication on gold coinage, providing a full description of the design (obverse and reverse) of each coin; the mint that struck the coin; the diameter and weight of the coin (metric measure); the percent of gold content; the number of coins in each issue with proof coins noted separately; the total troy-ounce weight of each issue; the total troy-ounce weight of gold used by each country; the total face value of all coins issued by each country; and the total weight of all gold used in coins worldwide. I have no doubt that *Modern Gold Coinage* is the most helpful gold coin list I have found.

22

Other Investment Information

There are other ways you can invest in precious metals, such as the futures market, South African gold shares, or North American gold shares. These ventures do not necessarily involve the physical possession of the precious metals. Of course, you can buy certificates for gold or silver bullion or coins that effectively give you title to ownership of the precious metals, without ever having to put up with the storage, security, handling, and insurance problems related to physical possession.

Any venture into this highly speculative field, however, requires (at the very least) consultations with reputable experts in the chosen field. Your banker, the firms mentioned in this book, or investment advisors recommended as established businesspeople can be consulted for a start.

It is recommended that you read up on such subjects by visiting the library or purchasing an authoritative book. There are several on the market. Two that I found most interesting are *How You Can Profit from Gold* by James E. Sinclair and Harry Schultz, published by Arlington House, and *How to Invest in Gold* by Peter C. Cavelti, published by Follett.

Part Three

Your Silver

23

Silver Trading

The California Gold Rush of 1849 has been eclipsed by the Great Precious Metals Scrap Drive of this century. The decade of the seventies ended and the eighties started with frenzied precious-metals trading. This phenomenon was spearheaded by a new factor in precious-metals dealing—the American public. For as long as many Americans could remember, a private citizen was prohibited from owning gold in specific forms. Our currency was backed by silver, and when silver was abandoned, only promises. Your watch, your ring, or even your tooth filling could be gold, but your pocket coins had to be silver (or copper-nickel). American gold coins from our glorious past could be possessed by collectors only. With the accent off gold, silver became the center of attention of American coin collectors. Silver coins abounded, while United States gold coins were few in comparison.

A prelude to the renewed interest in precious metals in the 1970s occurred in the late 1960s. The American public got a slight taste of things to come, and they didn't even realize they were dealing in precious metals. In 1965, the alloy used to make silver coins was changed, and no silver was actually used thereafter. All the public knew was that their old silver coins—which actually contained silver—could be sold for 25 percent over face value. The buyers were not speculators, but private smelters who illegally melted the coins.

The bubble burst, however, when the spot silver price dropped too low to keep the surreptitious smelters burning. Now the public is at it again, and many think they know what they're doing. They walk into a precious-metals-buying operation, often a hastily set up storefront business, sell their scrap, and walk out. They think they've cleaned up when, in reality, they've been taken to the cleaner's. Some of these dealers are honest, some dishonest, and some may be a little of both.

If you plan on selling your family-heirloom silverware, your high school ring, or maybe that set of dentures (with the gold tooth) left in your bottom bureau drawer for the last twenty years, the same rule applies to this transaction as to any business dealing: *know what you're doing.* An honest dealer can make a mistake, or a crooked dealer can take advantage of you, but if you know what you're doing, you can catch it. Remember, all you need to deal intelligently in precious metals is this book, a pocket calculator, and a pencil.

The first rule is *accept cash only* as payment for your precious metal. If the buyer is at a long-established location, you can probably safely accept a check. Quite often, however, such dealers prefer to pay by cash. Many municipalities and some states have passed laws requiring dealers to pay for purchases by check.

For example, an out-of-town buyer set up shop in a rented storefront on the main street of a city. He advertised lavishly in the local paper, paid for many purchases by check, closed up shop after a few days, and left town. His checks bounced. To this date, the dealer has not been found. Less than a block away, in either direction, were established dealers. Yet, many persons sold their scrap to the storefront operator. Proper licensing laws would have protected the public from such a dealer.

Below is an example of the daily silver and gold spot metal prices as found in a newspaper.

Silver and gold quotations

United Press International

SILVER

NEW YORK — Handy and Harman Wednesday quoted silver at $13.60 per fine ounce up $0.05.

Engelhard quoted a silver base price of $13.50 up $0.05 and a price for fabricated silver of $14.28 up $0.052.

GOLD

Foreign and domestic gold prices in dollars per troy ounce Wednesday.

NEW YORK — Handy and Harman 506.50 up 4.10.

Engelhard, base price for refining settling and unfabricated gold 507.75 up 4.10 per troy ounce. Selling price, fabricated gold 525.52 up 4.24 per troy ounce.

LONDON — Morning fixing 506.75 up 4.35. Afternoon fixing 506.50 up 4.10.

PARIS — (Free market) 527.61 up 9.97.

FRANKFURT — 509.77 up 12.86.
ZURICH — 507.50 up 5.00.

Over The Counter

	BID	ASK
Amadac Industries	½	⅞
Berkshire Gas	14	15½
Delhi Chemicals	1¼	1¾
Environment One	¼	½
Flah's Inc.	1½	2¼
Hamilton Digital	3½	4¼
Key Banks Inc.	11⅛	11⅝
Keystone Foods	11¾	12½
Mechanical Tech.	10¾	11¾
Nat. Micronetics	5¼	6
Schen. Trust	26	27
Tobin Packing Co.	2¾	3¾
United Bank Corp.	18¼	19
United Telecom.	10½	12

(Courtesy of Shearson, Loeb, Rhoades Inc. and First Albany Corp.)

While the silver and gold prices are not presented in exactly this format in every publication, the basic information is here. For example, silver is quoted in the newspaper at $13.60 from Handy and Harman, and $13.50 from Engelhard. This variance illustrates the fact that spot prices are not always the same at each center. Silver is shown to be up $0.05 at both refiners.

Gold is quoted at $506.50 from Handy and Harman, $507.75 from Engelhard; the London Market shows the morning fixing at $506.75, with the afternoon fixing at $506.50. Paris quotes $527.61, Frankfurt sets $509.77, and Zurich quotes $507.50. You can disregard the quote on fabricated precious metal, which is for industrial use.

There are many things to know, other than how to evaluate the precious metal content of your posessions, to deal with a gatherer intelligently. If you are selling a piece of jewelry with a gemstone in it, you must take into account the stone's value. It can be worth more than the precious metal in the item you are trading. Many buyers have accumulated a supply of precious stones that overzealous sellers have completely neglected in their frenzy to profit on the metal.

We have discussed what your "tools" are as the seller. The buyer's tool you need most to be concerned about is the scale. Some buyers use postal scales, food scales, and a wide assortment of "oddball" equipment. Be sure the buyer's scale is graduated by the (troy) ounce, pennyweight, grain, or metric gram. Food scales and other common weighing devices used in a merchant's daily transactions are avoirdupois weight, which is quite different from troy weight used in precious-metals dealings.

Many state, county, and/or municipal governments require that scales used in precious-metals operations be approved and inspected. Find out if there are laws or regulations that apply in your area. If there are, look for a seal of inspection on the scale. The official who inspects such scales holds them to strict standards. This is for the protection of the public. If you don't find an official seal of inspection on the scale of the buyer you are contemplating doing business with, ask that person to point it out. An honest dealer shouldn't get upset by any reasonable request or question. If the dealer is uncooperative, there are many other dealers to choose from. Find a dealer who will trade on a completely open basis.

To determine whether there are laws applying to inspection of

Sartorius digital scales appropriate for precious-metal weighings. From left to right: models 1202 GS, 1203 GS, and 3802 GS. (Photograph courtesy of Brinkmann Instruments, Inc.)

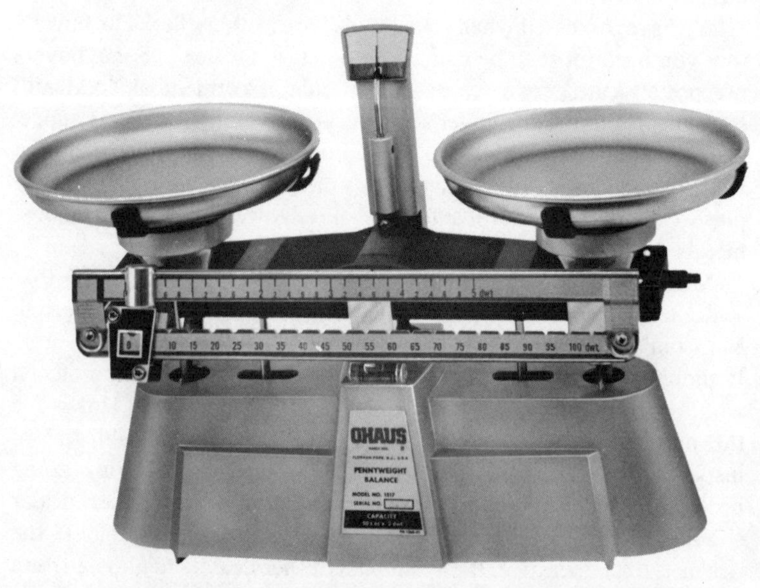

The more traditional mechanical scale, Ohaus scale, model 1517. (Photograph courtesy of Ohaus Scale Corp.)

scales in your area, call the offices of your local city or county government. The information desk can either give you the necessary information or direct you to the appropriate agency. For example, New York State requires approved scales and periodic inspections by trained experts. At this writing, ten scales are approved for precious-metals businesses by the state. Of the ten approved scales, two brands are conditionally approved subject to "proper maintenance." Of the scales approved unconditionally, two are manufactured by Ohaus Scale Corporation; and four are manufactured by Brinkmann Instruments, Inc. (Sartorius). The Ohaus scales have the more traditional look you might expect of scales, while the Sartorius models have microprocessor technology with digital readout. Both manufacturers' scales are pictured here. The Ohaus Model #1517 has a capacity of 1000 pennyweights, with a sensitivity correct to 0.2 pennyweight. The Brinkmann Sartorius Model #1202 has a capacity of 200 pennyweights with a precision of plus or minus 0.005 pennyweight. Model #1203 has a capacity of 2500 pennyweights, with a precision of plus or minus 0.05 pennyweight. Model #3802 has a capacity of 5200 pennyweights, with a precision of plus or minus 0.05 pennyweight. Ohaus and Sartorius both meet the New York State-approved tolerances, which is 0.05 percent of the test load.

In New York State, inspection and enforcement of scales used in precious-metals operations (and for any other weighing or measuring device used in commerce) is the duty of the county or municipal director of weights and measures. A replica of an official seal of inspection is reproduced here. Most states and most foreign countries

JAN.	1979	1981	1983	1985	JULY
FEB.	BUREAU of WEIGHTS and MEASURES				AUG.
MAR.					SEPT.
APR.	do not			detach	OCT.
MAY					NOV.
JUNE	SEALED COUNTY OF SARATOGA				DEC.
DAVID W. SEXTON DIRECTOR					

have regulations similiar to New York State. Find out for your own protection.

If you plan on dealing with a buyer and you *do* live in an area where approval and/or inspection of scales is required, it would be advisable to look for a dealer with approved scales. Many buyers have scales resurrected from attics and cellars, dusted off and put to use. Even though these dealers have had ample time and money to purchase the proper equipment, they are happy with their old scale. And for good reason. It should be unnecessary for me to explain what a set of inaccurate (or crooked) scales could do to your profits.

Virtually every calculation and chart is based on troy weight (with the exception of the metric measurement, gram). Avoirdupois weight is shown in comparison only on the chart "Measurements and Weights Used with Precious Metals." (One exception is made in the use of avoirdupois weight. The section "The United States Copper Penny" shows avoirdupois weight because the penny is a nonprecious metal coin, and, as such, doesn't use the troy system.

The abbreviations for each weight or measure are as follows: pound (lb.); ounce (oz.); pennyweight (dwt.); grain (gr.); metric gram (g.); millimeter (mm.). Remember: don't confuse grain (gr.) with metric gram (g.)

The troy-weight designation *pennyweight* is very confusing if you try to relate it to the United States one-cent (penny) coin. The penny is made of copper, which is not a precious metal. In dealings with this coin, the avoirdupois weight system is used, not troy weights. Avoirdupois weight, which is used for nonprecious metals such as copper, does not have the designation *pennyweight.*

24

Photographic Silver

When you think of silver, you think of a piece of jewelry, a coin, or houseware. With the new attention directed to precious metals, more people are probably aware of the uses of silver in photographic materials. Still, how many people know that photographic materials account for 42 percent of the total silver used in the United States, or that Eastman Kodak is the largest single industrial user of silver in the world? With the increased prices, recovery of silver in used materials or in outdated film and photographic paper is an important source for silver supplies. Kodak recovered twenty million ounces in 1979, almost thirty percent of the seventy million ounces used.

Photographic studios, film-processing plants, small photo shops, in fact, all firms, no matter how small, which use photographic supplies to any degree, can benefit from new systems for silver recovery that have been developed. The value of the solutions saved, which were once discarded, can pay for the silver recovery equipment within a short time, if it is used properly. Further, the environment benefits, since millions of ounces of pollutants are now recycled, rather than dumped down the drain. The cost of industrial-pollution control often adds millions of dollars to the cost of products. The silver-recovery process is especially noteworthy because it is a pollution control that brings in revenue. Units of varying sizes are available that will accommodate the needs of anyone from the small shop owner up to the large mail-order processors. Detailed information is available from Kodak representatives. If you are now throwing your used solutions out, you are throwing out money—silver money.

25

Sterling Silver and Scrap Silver

It somehow doesn't seem appropriate to group sterling silver with silver scrap. But it is sadly true that, with few exceptions, almost every item made out of precious metal will ultimately be relegated to the smelting furnace. A beautiful necklace, passing through many hands, ends up in the melting pot. And from the resulting ingot, the coins of a new nation might be formed, or the bar may be put into solution over many sheets of photographic paper. Whatever the use, it will probably experience many such reincarnations. More than ever before, silver is being saved.

If you contemplate selling any silver piece for its scrap-silver value, evaluate it first. In many cases, the piece you want to dispose of has a greater value than the intrinsic silver content. If care has been taken to create a piece of jewelry, a set of silverware, and so on, it is probable that the piece has value as a work of art. Many gatherers purchased beautiful silver pieces for the scrap-silver value alone, with no intention of having the piece refined. They were polished and resold for prices well over the value of the silver weight.

Of course, the same can be said for silver-plated items. Gatherers who offer to buy silver plate are more apt to be interested in the item's artistic value, rather than the relatively small amount of silver that could be reclaimed by smelting. When only one-hundredth to one-five-hundredth of the total weight is sterling silver, the worth of an item is more likely to be aesthetic rather than intrinsic.

Items marked "silver," "solid silver," "sterling," or "sterling silver" must all be an alloy of at least 925 fine (with allowable deviations). Items marked "coin" or "coin silver" must be an alloy of at least 900 fine (with allowable deviations). Silver-plated items must be plainly marked as such, and the designations "sterling" or "coin" cannot appear anywhere on such an item, either alone or with the words "silver plate" or "plate."

If you are determined to sell, or an expert tells you your piece has little aesthetic value, you should shop for the best deal. The deals gatherers offer often vary considerably, so call the different dealers and write down the price they pay for sterling silver. The different deals may be presented as "so much an ounce," "so much a pound," or "so much a pennyweight." Whatever the deal is, it should be evaluated, using the methods described in this book, in the quiet of your own home.

Many silver pieces will have plaster, lead, or another heavy foreign material filling a hollow area, acting as a baseweight or counterbalance. The dealer may make this known to you and (quite often) will tell you he must *estimate* the weight of the added foreign material so he can deduct the extra weight from the total. If you are determined to sell the item, do not accept this method of deducting the added weight of the counterbalance material. Tell the dealer you want to take the piece back home to remove the foreign material by prying or cutting it off. You may find that the dealer does not want to destroy the shape or design of the silver piece. This reluctance would be a good indication that the dealer was interested in the piece for its artistic worth rather than the scrap silver. There is nothing wrong with that, *if* the buyer is willing to buy the piece for an equitable price, one reflecting its value as an art object.

Many beautiful sterling silver pieces were sold in the frenzied tradings of the Great Precious Metals Scrap Drive. Whether they found their way to the smelter's furnace or were reserved for the display cases of the purchasers is a moot point. The fact remains, the prices paid for much of the gold and silver scrap were unreasonably low. Only because of an unknowledgeable public, caught in a trading frenzy that may never again be equaled in scope, were the gatherers able to garner so much for so little cash outlay.

Silverware was probably one of the more heavily traded items. Sets that had been in families for generations, some of them collectors' pieces of rare quality and antique value, were sold by the set or individual piece.

Payment was offered by the weight and by the piece. The public was most often taken advantage of when selling a silverware knife. Knives require a stainless steel blade, rather than one made of silver, to provide a good cutting edge. Silver is too soft to hold a sharp edge, and could more easily be bent. Further, the knife handle usually has a filler material. With these foreign materials in the knives, some gatherers had a field day. The seller was told the knife had these non-silver materials, and after weighing the knife it would be necessary to *estimate* the weight of the silver content, to make allowances for the non-silver content. Any gatherer smart enough to know that knives were not all sterling, should also be smart enough to know the gross and net weights of the different knife designs and patterns. I watched one gatherer consult a paper he had kept face down under his desk. It was a list of many silverware patterns with each name followed by the gross and net weights of that specific pattern. That gatherer didn't have to estimate, he had the exact figures to consult. All he had to do was check the gross weight of the knife he had just weighed against the gross weight of that particular pattern on his list; if the weights were identical, he knew the exact net weight.

For example, the oldest and largest selling design for silverware in the entire world is the Chantilly pattern by Gorham Textron. Originally introduced in 1895, it is still a top seller. A photograph of a place setting of Chantilly is featured on page 88.

The Chantilly pattern, in continuous production since 1895, is the most popular silverware ever produced. (Photograph courtesy of Gorham Silverware.)

The knife in this pattern, which takes more intricate handwork than any other piece, requires high craftsmanship in joining the blade parts and handle. It has a silver weight of 0.51 oz. Obviously, this information is not known to the man on the street, but no gatherer worth his salt should be without it. Still, many gatherers tell unsuspecting sellers that the weight of the silver in the knife will have to be estimated.

A simple method of calculating the value of your sterling-silver scrap (shown below) is helpful in dealing with precious-metals buyers.*

Sterling silver is 925 fine; therefore, 1 troy oz. of sterling silver (in the troy system, 12 oz. equals 1 lb.) has 11.1 oz. of fine silver. Using as an example a dealer's offer of $200 per pound for silver scrap, you would use the following procedure in calculating the fairness of his price.

1. Since the dealer is offering a per-pound price and the spot price is based on an ounce, you must first calculate what the dealer's offer is per ounce of fine silver.

$200 per lb.
÷ 11.1 oz. fine gold per pound sterling silver
= $18.01

The dealer is paying $18.01 per oz. of fine sterling; but the spot price is again at $35.

2. Therefore the dealer's profit is:

$35.00 spot price
− $18.01 dealer's offer
$16.99 dealer's gross profit

Subtract the dealer's refining costs:

$16.99 dealer's gross profit
− $ 0.19 refining costs per ounce
$16.80 dealer's net profit per ounce

* Remember that knives are an exception, as discussed above.

Remember that in addition to pounds and ounces, the dealer's scale can also be calibrated in pennyweight and grams.

If the scale is in pennyweight, the total can be converted to ounces by dividing the pennyweights by 20 (20 dwt. equals 1 oz.). For example, 27.752 dwt. equals 1.3876 oz. (27.752 dwt. ÷ 20 = 1.3876 oz.).

If the scale is in grams, the total can be converted to ounces by dividing by 31.104 (31.104 grams equals 1 oz.). For example, 43.1599 g. equals 1.3876 oz. (43.1599 g. ÷ 31.104 = 1.3876 oz.). Your notepad, with all the relevant information and calculations, should look like the following (with the gatherer's advertisement which offers $200 per pound for sterling-silver scrap, *all* the calculations can be done at home or at your office):

Spot Silver Price $35.
Sterling Silver fineness 925
refining loss, costs 19¢ per ounce

(use 11.1 for 'per pound'; use .925 for 'per ounce')

Dealer's offer (from ad) $200 lb.

11.1) $200.00 = $ 18.01

$35.00
− 0.19
$ 34.81
− 18.01
$ 16.80
(dealer's profit)

weight of my silver
(on buyer's scale)

1.3876 lbs.
× $ 200
$ 277.52 total paid

Copy all these calculations on your pad, even though you use a calculator, for an immediate reference at the gatherer's, and a later record.

26

Contemporary Silver Coins Worldwide

The following chart includes all silver coins issued as legal tender worldwide in 1979. This will give you an idea of contemporary silver coinage and its fine-silver value.

The Silver Institute, 1001 Connecticut Ave., N.W., Washington, D.C. 20036, publishes *Modern Silver Coinage* each year. This is a complete list of all silver coins issued worldwide. The cost is $10.00, and the valuable information includes a full description of each coin; the mint that struck the coin; the diameter of the coin and its weight in grams; the percentage of silver content; the exact number of coins in each issue; the number of "proof" coins in each issue; the total troy-ounce weight of the issue; the total troy-ounce weight of silver used by each country; the total face value of all coins issued by each country; and the total weight of all silver used in coins worldwide. There is no question that *Modern Silver Coinage* is one of the most useful coin lists ever published.

*Worldwide Silver Coinage 1979**

COUNTRY	COIN	COIN WEIGHT	FINE-NESS	WEIGHT OF FINE SILVER	NUMBER OF COINS ISSUED
Afghanistan	500 Afghanis	1.1253 oz.	925	1.0409 oz.	1,351
	250 Afghanis	0.9092 oz.	925	0.8410 oz.	1,364
Andorra	2,000 Pesetas	0.8038 oz.	925	0.7435 oz.	1,000
Austria	* 100 Schillings	0.7716 oz.	640	0.4938 oz.	2,000,000
* *(4 issues—*	* 100 Schillings	0.7716 oz.	640	0.4938 oz.	2,000,000
4 different designs)	* 100 Schillings	0.7716 oz.	640	0.4938 oz.	2,000,000
	* 100 Schillings	0.7716 oz.	640	0.4938 oz.	1,800,000
Bahamas	25 Dollars	1.2018 oz.	925	1.1116 oz.	3,002
* *(3 issues—*	5 Dollars	1.3542 oz.	925	1.2526 oz.	2,053
3 different designs)	2 Dollars	0.9584 oz.	925	0.8865 oz.	2,053
	1 Dollar	0.5832 oz.	800	0.4666 oz.	2,053
	50 Cents	0.3334 oz.	800	0.2667 oz.	2,053
	* 10 Dollars	1.4583 oz.	500	0.7292 oz.	10,939
	* 10 Dollars	1.4583 oz.	500	0.7292 oz.	5,073
	* 10 Dollars	1.4583 oz.	500	0.7292 oz.	18,840
Barbados	10 Dollars	1.2217 oz.	925	1.1301 oz.	6,534
	5 Dollars	1.0320 oz.	800	0.8256 oz.	4,126
Belize	10 Dollars	0.9584 oz.	925	0.8865 oz.	2,445
	5 Dollars	0.8584 oz.	925	0.7940 oz.	2,445
	1 Dollar	0.6308 oz.	925	0.5835 oz.	2,445
	50 Cents	0.3373 oz.	925	0.3120 oz.	2,445

* (Courtesy of the Silver Institute, and adapted from its publication *Modern Silver Coinage.*)

	25 Cents	0.2103 oz.	925	0.1945 oz.	2,445
	10 Cents	0.0891 oz.	925	0.0824 oz.	2,445
	5 Cents	0.1379 oz.	925	0.1276 oz.	2,445
	1 Cent	0.0981 oz.	925	0.0907 oz.	2,445
Benin	1,000 Francs	1.6557 oz.	999	1.6541 oz.	600
	500 Francs	0.8263 oz.	999	0.8254 oz.	500
	200 Francs	0.3311 oz.	999	0.3308 oz.	200
	100 Francs	0.1640 oz.	999	0.1638 oz.	200
Botswana	10 Pula	1.1253 oz.	925	1.0409 oz.	549
	5 Pula	0.9092 oz.	925	0.8410 oz.	1,103
	5 Pula	0.9092 oz.	500	0.8546 oz.	557
British Virgin Islands	5 Dollars	1.1291 oz.	925	1.1937 oz.	5,304
	1 Dollar	0.8070 oz.	925	0.7464 oz.	5,304
Brunei	10 Dollars	0.9092 oz.	925	0.8410 oz.	411
Bulgaria	20 Leva	0.7009 oz.	500	0.3504 oz.	35,000
	20 Leva	1.0288 oz.	900	0.9259 oz.	15,000
	10 Leva	0.4501 oz.	500	0.2251 oz.	35,000
	10 Leva	0.7716 oz.	900	0.6944 oz.	15,000
	5 Leva	0.6591 oz.	500	0.3295 oz.	50,000
Canada	1 Dollar	0.7500 oz.	500	0.3750 oz.	826,695
Cayman Islands	50 Dollars	2.0833 oz.	925	1.9271 oz.	775
* *(6 issues—*	* 25 Dollars	1.6667 oz.	925	1.5417 oz.	2,975
6 different designs)	* 25 Dollars	1.6667 oz.	925	1.5417 oz.	2,896
	* 25 Dollars	1.6667 oz.	925	1.5417 oz.	2,756
	* 25 Dollars	1.6667 oz.	925	1.5417 oz.	3,503
	* 25 Dollars	1.6667 oz.	925	1.5417 oz.	3,394
	* 25 Dollars	1.6667 oz.	925	1.5417 oz.	3,660

Worldwide Silver Coinage 1979—Continued

COUNTRY	COIN	COIN WEIGHT	FINE-NESS	WEIGHT OF FINE SILVER	NUMBER OF COINS ISSUED
Chile	10 Pesos	1.4468 oz.	999.9	1.4466 oz.	650
Cook Islands	5 Dollars	0.9044 oz.	500	0.4522 oz.	11,362
Costa Rica	* 100 Colones	1.1253 oz.	925	1.0409 oz.	100,000
* *(3 issues—2 designs,*	* 100 Colones	1.1253 oz.	500	0.5626 oz.	17
2 degrees of fineness)	* 100 Colones	1.1253 oz.	925	1.0409 oz.	140
	50 Colones	0.9092 oz.	500	0.4546 oz.	14
	50 Colones	0.9092 oz.	925	0.8410 oz.	140
	25 Colones	1.7329 oz.	999	1.7312 oz.	100
	20 Colones	1.3863 oz.	999	1.3849 oz.	100
	10 Colones	0.6932 oz.	999	0.6925 oz.	50
	5 Colones	0.3466 oz.	999	0.3462 oz.	50
	2 Colones	0.1386 oz.	999	0.1384 oz.	50
Cuba	20 Pesos	0.8359 oz.	925	0.7732 oz.	20,000
Cyprus	500 Mils	0.4546 oz.	925	0.4205 oz.	3,600
Czechoslovakia	100 Korun	0.4823 oz.	700	0.3376 oz.	80,000
	50 Korun	0.4180 oz.	700	0.2926 oz.	100,000
Dominica	* 20 Dollars	1.3182 oz.	925	1.2193 oz.	1,300
* *(2 issues—same design)*	* 20 Dollars	1.3182 oz.	925	1.2193 oz.	251
** *(3 issues—2 designs)*	** 10 Dollars	0.6591 oz.	925	0.6096 oz.	750
	** 10 Dollars	0.6591 oz.	925	0.6096 oz.	5,700
	** 10 Dollars	0.6591 oz.	925	0.6096 oz.	251
Dominican Republic	25 Pesos	2.0898 oz.	925	1.9330 oz.	9,000

Egypt	* 1 Pound	0.4823 oz.	720	0.3472 oz.	100,000
(6 issues—6 designs)	* 1 Pound	0.4823 oz.	720	0.3472 oz.	100,000
	* 1 Pound	0.4823 oz.	720	0.3472 oz.	100,000
	* 1 Pound	0.4823 oz.	720	0.3472 oz.	100,000
	* 1 Pound	0.4823 oz.	720	0.3472 oz.	100,000
	* 1 Pound	0.4823 oz.	720	0.3472 oz.	50,000
El Salvador	1 Colon	0.0739 oz.	999.9	0.0739 oz.	2,000
Ethiopia	25 Birr	1.1253 oz.	925	1.0409 oz.	6,934
	10 Birr	0.9092 oz.	925	0.8410 oz.	7,259
Equatorial Guinea	2,000 Ekuele	0.9999 oz.	927	0.9269 oz.	2,500
* *(2 issues—2 designs)*	* 2,000 Ekuele	1.3783 oz.	924.9	1.2748 oz.	88
	* 2,000 Ekuele	1.3783 oz.	924.9	1.2748 oz.	134
	1,000 Ekuele	0.6890 oz.	924.9	0.6372 oz.	144
	200 Pesetas	1.2860 oz.	999	1.2847 oz.	250
	100 Pesetas	0.6430 oz.	999	0.6424 oz.	1,450
	75 Pesetas	0.4823 oz.	999	0.4818 oz.	50
Falkland Islands	10 Pounds	1.1253 oz.	925	1.0409 oz.	6,720
	5 Pounds	0.9092 oz.	925	0.8410 oz.	6,864
	50 Pence	0.9092 oz.	925	0.8410 oz.	1,942
Fiji	25 Dollars	1.5622 oz.	925	1.4450 oz.	4
	20 Dollars	1.1253 oz.	925	1.0409 oz.	3,704
	10 Dollars	0.9092 oz.	925	0.8410 oz.	4,143
	1 Dollar	1.0095 oz.	925	0.9338 oz.	69
	50 Cents	0.5594 oz.	925	0.5175 oz.	69
	20 Cents	0.4019 oz.	925	0.3717 oz.	69
	10 Cents	0.2009 oz.	925	0.1859 oz.	69
	5 Cents	0.1003 oz.	925	0.0928 oz.	69
	2 Cents	0.1382 oz.	925	0.1279 oz.	69
	1 Cent	0.0691 oz.	925	0.0639 oz.	69

Worldwide Silver Coinage 1979—Continued

COUNTRY	COIN	COIN WEIGHT	FINE-NESS	WEIGHT OF FINE SILVER	NUMBER OF COINS ISSUED
Finland	25 Markkaa	0.8456 oz.	500	0.4228 oz.	300,000
France	50 Francs	1.9290 oz.	950	1.8326 oz.	2,266
* *(2 issues—*	* 50 Francs	0.9645 oz.	900	0.8681 oz.	8,656,920
comparable designs)	* 50 Francs	0.9645 oz.	900	0.8681 oz.	40,500
	10 Francs	0.7330 oz.	950	0.6964 oz.	716
	5 Francs	0.7330 oz.	950	0.6964 oz.	616
	2 Francs	0.5723 oz.	950	0.5437 oz.	1,266
	1 Franc	0.4405 oz.	950	0.4184 oz.	616
	½ Franc	0.3537 oz.	950	0.3360 oz.	616
	20 Centimes	0.3279 oz.	950	0.3115 oz.	616
	10 Centimes	0.2443 oz.	950	0.2321 oz.	616
	5 Centimes	0.1608 oz.	950	0.1527 oz.	616
	1 Centime	0.1415 oz.	950	0.1344 oz.	616
French Polynesia	100 Francs	0.7620 oz.	950	0.7239 oz.	357
	50 Francs	1.1413 oz.	950	1.0843 oz.	258
	20 Francs	0.7620 oz.	950	0.7239 oz.	257
	10 Francs	0.4565 oz.	950	0.4337 oz.	257
	5 Francs	0.8841 oz.	950	0.8399 oz.	257
	2 Francs	0.5626 oz.	950	0.5345 oz.	257
	1 Franc	0.3279 oz.	950	0.3115 oz.	257
Gambia	40 Dalasi	1.1253 oz.	925	1.0409 oz.	338
German Democratic Republic	20 Marks	0.6719 oz.	500	0.3360 oz.	45,000
	10 Marks	0.5466 oz.	500	0.2733 oz.	55,000

German Federal Republic	* 5 Marks	0.3601 oz.	625	0.2251 oz.	8,000,000
* *(2 issues—2 designs)*	* 5 Marks	0.3601 oz.	625	0.2251 oz.	8,350,000
Gibraltar	25 Pence (1 Crown)	0.9092 oz.	925	0.8410 oz.	881
Great Britain	4 Pence	0.0608 oz.	925	0.0562 oz.	1,188
	3 Pence	0.0453 oz.	925	0.0419 oz.	1,294
	2 Pence	0.0302 oz.	925	0.0280 oz.	1,188
	1 Pence	0.0151 oz.	925	0.0140 oz.	1,188
Greece	500 Drachmas	0.4180 oz.	900	0.3762 oz.	18,000
	100 Drachmas	0.4180 oz.	650	0.2717 oz.	25,000
Guernsey	25 Pence (1 Crown)	0.9092 oz.	925	0.8410 oz.	1,256
Guinea	500 Francs	0.9295 oz.	999	0.9285 oz.	2,950
* *(2 issues—2 designs)*	* 500 Sylis	1.3503 oz.	925	1.2490 oz.	100
	* 500 Sylis	1.3503 oz.	925	1.2490 oz.	100
Guyana	10 Dollars	1.3953 oz.	925	1.2907 oz.	2,665
	5 Dollars	1.2140 oz.	500	0.6070 oz.	2,665
Haiti	* 50 Gourdes	0.6912 oz.	925	0.6394 oz.	350
* *(2 issues—2 designs)*	* 50 Gourdes	0.6912 oz.	925	0.6394 oz.	250
	25 Gourdes	0.2694 oz.	925	0.2492 oz.	500
	10 Gourdes	0.5127 oz.	999	1.5111 oz.	1,400
Hungary	200 Forint	0.9002 oz.	640	0.5761 oz.	30,000
	200 Forint	1.8004 oz.	640	1.1523 oz.	2,500
	200 Forint	.7073 oz.	640	0.4527 oz.	20,000
	200 Forint	1.4146 oz.	640	0.9053 oz.	2,500
Isle of Man	* 1 Crown	0.9092 oz.	925	0.8410 oz.	100,000
* *(6 issues—6 designs)*	* 1 Crown	0.9092 oz.	925	0.8410 oz.	35,000

Worldwide Silver Coinage 1979—Continued

COUNTRY	COIN	COIN WEIGHT	FINE-NESS	WEIGHT OF FINE SILVER	NUMBER OF COINS ISSUED
	* 1 Crown	0.9092 oz.	925	0.8410 oz.	35,000
	* 1 Crown	0.9092 oz.	925	0.8410 oz.	35,000
	* 1 Crown	0.9092 oz.	925	0.8410 oz.	35,000
	* 1 Crown	0.9092 oz.	925	0.8410 oz.	35,000
Iraq	1 Dinar	0.9967 oz.	900	0.8970 oz.	5,000
Israel	100 Lirot	0.6430 oz.	500	0.3215 oz.	53,000
* *(2 issues—same design)*	* 50 Lirot	0.6430 oz.	500	0.3215 oz.	16,200
	* 50 Lirot	0.6430 oz.	500	0.3215 oz.	24,200
Jamaica	25 Dollars	4.3724 oz.	925	4.0445 oz.	17,646
	25 Dollars	4.3750 oz.	925	4.0468 oz.	2,633
	10 Dollars	1.3542 oz.	925	1.2526 oz.	8,308
	10 Dollars	0.7501 oz.	925	0.6938 oz.	11,900
	10 Dollars	1.5001 oz.	925	1.3876 oz.	2,000
Jordan	3 Dinars	1.1253 oz.	925	0.0409 oz.	133
	2½ Dinars	0.9092 oz.	925	0.8410 oz.	114
Kenya	200 Shillings	0.9092 oz.	925	0.8410 oz.	9,500
Kiribati	5 Dollars	0.9092 oz.	500	0.4546 oz.	1,980
	5 Dollars	0.9092 oz.	925	0.8410 oz.	4,877
Lesotho	15 Maloti	1.0799 oz.	925	0.9991 oz.	500
	15 Maloti	1.0811 oz.	925	0.9999 oz.	1,000
	10 Maloti	0.9092 oz.	925	0.8410 oz.	7,902
	10 Maloti	0.9092 oz.	500	0.4546 oz.	2,000
Liberia	5 Dollars	1.1458 oz.	900	1.0312 oz.	1,857

Macao	100 Patacas	0.9092 oz.	925	0.8410 oz.	5,500
Madagascar	20 Ariary	0.3858 oz.	925	0.3569 oz.	1,598
	10 Ariary	0.2894 oz.	925	0.2676 oz.	1,598
Malawi	10 Kwacha	1.1253 oz.	925	1.0409 oz.	5,734
	10 Kwacha	0.9092 oz.	925	0.8410 oz.	307
	5 Kwacha	0.9092 oz.	925	0.8410 oz.	5,732
Malaysia	25 Ringgits	1.1253 oz.	925	1.0409 oz.	251
	15 Ringgits	0.9092 oz.	925	0.8410 oz.	211
	1 Ringgit	0.5417 oz.	925	0.5011 oz.	8,000
Maldives	25 Rupees	0.9092 oz.	925	0.8410 oz.	2,000
	5 Rupees	0.6109 oz.	925	0.5650 oz.	1,887
Malta	* 1 Pound	0.1820 oz.	925	0.1683 oz.	50,000
* *(2 issues—1 design)*	* 1 Pound	0.1820 oz.	925	0.1683 oz.	7,871
Mauritius	50 Rupees	1.1253 oz.	500	0.5626 oz.	1,418
* *(2 issues—*	50 Rupees	1.1253 oz.	925	1.0409 oz.	150
2 reverse designs)	25 Rupees	0.9092 oz.	500	0.4546 oz.	1,397
	* 25 Rupees	0.9092 oz.	925	0.8410 oz.	151
	* 25 Rupees	0.9092 oz.	925	0.8410 oz.	3,000
Mexico	100 Pesos	0.8930 oz.	720	0.6430 oz.	783,500
	Onzo Troy	1.0811 oz.	925	1.0000 oz.	4,508,000
	(This is a "bullion coin")				
Mongolia	50 Tukhrik	1.1253 oz.	925	1.0409 oz.	217
	25 Tukhrik	0.9092 oz.	925	0.8410 oz.	216
Morocco	* 50 Dirhams	1.6982 oz.	925	1.5708 oz.	20
* *(3 issues—3 designs)*	* 50 Dirhams	1.6982 oz.	925	1.5708 oz.	20
** *(3 issues—3 designs)*	* 50 Dirhams	1.6982 oz.	925	1.5708 oz.	25
	** 50 Dirhams	1.1368 oz.	925	1.0516 oz.	5,500

Worldwide Silver Coinage 1979—Continued

COUNTRY	COIN	COIN WEIGHT	FINE-NESS	WEIGHT OF FINE SILVER	NUMBER OF COINS ISSUED
	** 50 Dirhams	1.1368 oz.	925	1.0516 oz.	5,700
	** 50 Dirhams	1.1368 oz.	925	1.0516 oz.	3,478
Nepal	50 Rupees	0.8038 oz.	600	0.4823 oz.	261
	50 Rupees	1.1253 oz.	500	0.5626 oz.	40
	50 Rupees	1.1253 oz.	925	1.0409 oz.	140
	25 Rupees	0.9092 oz.	500	0.4546 oz.	10
	25 Rupees	0.9092 oz.	925	0.8410 oz.	140
New Caledonia	100 Francs	0.7620 oz.	950	0.7239 oz.	357
	50 Francs	1.1413 oz.	950	1.0843 oz.	257
	20 Francs	0.7620 oz.	950	0.7239 oz.	257
	10 Francs	0.4565 oz.	950	0.4337 oz.	257
	5 Francs	0.8841 oz.	950	0.8399 oz.	257
	2 Francs	0.5626 oz.	950	0.5345 oz.	257
	1 Franc	0.3279 oz.	950	0.3115 oz.	257
New Hebrides	50 Francs	1.1413 oz.	950	1.0843 oz.	257
	20 Francs	0.7620 oz.	950	0.7239 oz.	257
	10 Francs	0.4565 oz.	950	0.4337 oz.	257
	5 Francs	0.8841 oz.	950	0.8399 oz.	257
	2 Francs	0.5626 oz.	950	0.5345 oz.	257
	1 Franc	0.3279 oz.	950	0.3115 oz.	257
New Zealand	1 Dollar	0.8750 oz.	925	0.8094 oz.	35,000
Oman	5 Rials	1.1253 oz.	925	1.0409 oz.	241
	2½ Rials	0.9092 oz.	925	0.8410 oz.	11
	1 Rial	0.4823 oz.	500	0.2411 oz.	100

Pakistan	150 Rupees	1.1253 oz.	925	1.0409 oz.	290
	100 Rupees	0.9092 oz.	925	0.8410 oz.	290
Panama	20 Balboas	4.1663 oz.	925	3.8538 oz.	15,000
	5 Balboas	1.1291 oz.	925	1.0444 oz.	5,949
	1 Balboa	0.8636 oz.	925	0.7988 oz.	7,160
Papua New Guinea	10 Kinas	1.3220 oz.	925	1.2526 oz.	5,595
	5 Kinas	0.8851 oz.	500	0.4425 oz.	4,260
Peru	5000 Soles	1.0812 oz.	925	1.0001 oz.	100,000
	1000 Soles	0.5006 oz.	500	0.2503 oz.	200,000
	200 Soles	0.7073 oz.	800	0.5658 oz.	3,000
Philippines	50 Piso	0.9002 oz.	925	0.8327 oz.	37,099
	25 Piso	0.8038 oz.	500	0.4019 oz.	17,093
Poland	200 Zlotych	0.5658 oz.	750	0.4244 oz.	8,000
* *(4 issues—4 designs)*	* 100 Zlotych	0.5305 oz.	625	0.3315 oz.	30,000
	* 100 Zlotych	0.5305 oz.	625	0.3315 oz.	30,000
	* 100 Zlotych	0.5305 oz.	625	0.3315 oz.	20,000
	* 100 Zlotych	0.5305 oz.	625	0.3315 oz.	20,000
San Marino	1000 Lire	0.4691 oz.	835	0.3916 oz.	125,000
	500 Lire	0.3537 oz.	835	0.2953 oz.	125,000
Sao Tome and Principe	* 250 Dobras	0.5594 oz.	925	0.5175 oz.	200
* *(5 issues—5 designs)*	* 250 Dobras	0.5594 oz.	925	0.5175 oz.	200
	* 250 Dobras	0.5594 oz.	925	0.5175 oz.	200
	* 250 Dobras	0.5594 oz.	925	0.5175 oz.	200
	* 250 Dobras	0.5594 oz.	925	0.5175 oz.	200
Seychelles	100 Rupees	1.1253 oz.	925	1.0409 oz.	18
	50 Rupees	0.9092 oz.	925	0.8410 oz.	219
Singapore	10 Dollars	0.9999 oz.	500	0.5000 oz.	180,500
	1 Dollar	0.5803 oz.	925	0.5368 oz.	8,000

Worldwide Silver Coinage 1979—Continued

COUNTRY	COIN	COIN WEIGHT	FINE-NESS	WEIGHT OF FINE SILVER	NUMBER OF COINS ISSUED
Solomon Islands	10 Dollars	1.3092 oz.	925	1.2110 oz.	4,670
	5 Dollars	0.8941 oz.	925	0.8270 oz.	2,845
Somalia	10 Shillings	0.9092 oz.	925	0.8410 oz.	3,250
South Africa	1 Rand	0.4823 oz.	800	0.3858 oz.	15,000
Saint Helena	25 Pence (1 Crown)	0.9092 oz.	925	0.8410 oz.	500
Sudan	10 Pounds	1.1253 oz.	925	1.0409 oz.	1,438
	5 Pounds	0.5626 oz.	925	0.5204 oz.	1,444
	5 Pounds	1.1253 oz.	925	1.0409 oz.	308
	2½ Pounds	0.9092 oz.	925	0.8410 oz.	297
Tanzania	50 Shillings	1.1253 oz.	500	0.5626 oz.	17
	50 Shillings	1.1253 oz.	925	1.0409 oz.	121
	25 Shillings	0.9092 oz.	500	0.4546 oz.	20
	25 Shillings	0.9092 oz.	925	0.8410 oz.	219
Thailand	300 Baht	0.7073 oz.	925	0.6543 oz.	20,000
	200 Baht	0.7073 oz.	925	0.6543 oz.	50,000
	100 Baht	1.1253 oz.	925	1.0409 oz.	182
	50 Baht	0.9092 oz.	925	0.8410 oz.	170
Tokelau	1 Dollar	0.8751 oz.	925	0.8095 oz.	5,500
Tonga	2 Pa'anga	1.2217 oz.	999	1.2205 oz.	850
	1 Pa'anga	0.6109 oz.	999	0.6102 oz.	850
Trinidad and Tobago	10 Dollars	1.1291 oz.	925	1.0444 oz.	5,144
	5 Dollars	0.9584 oz.	925	0.8865 oz.	4,055

Tunisia	10 Dinars	1.2217 oz.	900	1.0995 oz.	10,000
Turkey	* 150 Liras	0.2894 oz.	900	0.2604 oz.	12,500
(2 issues—same design, different dates)	* 150 Liras	0.2894 oz.	900	0.2604 oz.	12,500
Turks & Caicos Islands	* 25 Crowns	1.4062 oz.	925	1.3008 oz.	(note below)

* *Note: There are 10 issues of this coin—same denomination, size, and weight, each with a different design (10 designs). Number of coins in each issue as follows: 2,877; 2,797; 3,446; 2,842; 3,195; 592; 3,072; 3,061; 2,959; 3,001.*

	20 Crowns	1.2500 oz.	925	1.1562 oz.	1,579
	10 Crowns	0.9581 oz.	925	0.8862 oz.	14,733
Tuvalu	10 Dollars	1.1253 oz.	925	1.0409 oz.	1,500
	10 Dollars	1.1253 oz.	500	0.5626 oz.	2,560
Union of Soviet	* 10 Roubles	1.0710 oz.	900	0.9639 oz.	450,000
Socialist Republics	* 10 Roubles	1.0710 oz.	900	0.9639 oz.	450,000
* *(3 issues—3 designs)*	* 10 Roubles	1.0710 oz.	900	0.9639 oz.	450,000
** *(2 issues—2 designs)*	** 5 Roubles	0.5359 oz.	900	0.4823 oz.	450,000
	** 5 Roubles	0.5359 oz.	900	0.4823 oz.	450,000
United States of America	1 Dollar	0.7906 oz.	400	0.3162 oz.	312,268
	50 Cents	0.3697 oz.	400	0.1479 oz.	312,268
	25 Cents	0.1849 oz.	400	0.0739 oz.	312,268
Uruguay	5 Centesimos	0.2379 oz.	900	0.2141 oz.	202
	2 Centesimos	0.1672 oz.	900	0.1505 oz.	202
	1 Centesimo	0.1190 oz.	900	0.1071 oz.	202
Vatican City	1,000 Lire	0.4692 oz.	835	0.3918 oz.	200,000
Venezuela	50 Bolivares	1.1253 oz.	500	0.5626 oz.	41
Western Samoa	10 Dollars	0.9324 oz.	925	0.8624 oz.	5,000
	10 Dollars	0.5048 oz.	500	0.2524 oz.	3,000

Worldwide Silver Coinage 1979—Concluded

COUNTRY	COIN	COIN WEIGHT	FINE-NESS	WEIGHT OF FINE SILVER	NUMBER OF COINS ISSUED
Yemen Arab Republic	2 Riyals	0.8038 oz.	925	0.7435 oz.	2,000
	1 Riyal	0.3858 oz.	925	0.3569 oz.	2,000
Yugoslavia	400 Dinars	0.8038 oz.	925	0.7435 oz.	19,262
	350 Dinars	0.7234 oz.	925	0.6691 oz.	14,762
	300 Dinars	0.6430 oz.	925	0.5948 oz.	18,214
	250 Dinars	0.5626 oz.	925	0.5204 oz.	19,820
	200 Dinars	0.4823 oz.	925	0.4461 oz.	21,476
	150 Dinars	0.4019 oz.	925	0.3717 oz.	24,239
	100 Dinars	0.3215 oz.	925	0.2974 oz.	23,423
Zaire	5 Zaires	1.1253 oz.	925	1.0409 oz.	160
	2½ Zaires	0.9092 oz.	925	0.8410 oz.	160
Zambia	10 Kwacha	1.1253 oz.	925	1.0409 oz.	5,196
	5 Kwacha	0.9092 oz.	925	0.8410 oz.	5,385

27

Charts on United States and Canadian Silver Coins

The next sections provide data on United States and Canadian silver coins. The charts included show actual coin weight and the weight of fine silver in ounces, pennyweights, and grams. In the information preceding each chart, the method for calculating the value of the fine-silver content and converting pennyweights and grams to ounces is shown. Actually, the charts are laid out to show the gross weight and the actual weight of the silver content in the coins under consideration. With this information in hand, all that is necessary for analyzing the gatherer's offer for your scrap is the spot silver price. (Of course, make sure to check out the numismatic value of any silver coin before trading it solely for its fine silver content. For more on this theme, see section 20 above, "Numismatic Value.")

For example, assume you have $100 in War Nickels (face value). The chart for War Nickels shows the fine-silver content as 112.5 oz. To find the full value of the nickels, you multiply the spot price (assume it is $16 per oz. of fine silver) by 112.5: 112.5 oz. × $16 per oz. = $1800. The value of the coins is $1800.

If the dealer is offering $6 for every $1 in face value of the nickels, he would obviously be offering $600 for the $100 of War Nickels (100 × 6 = $600).

Before you can calculate the dealer's profit, you must subtract his refining costs and losses. As a rule, about 2 percent of fine silver is lost in the refining process.

1. 112.5 oz. fine silver
   × 0.02 loss
   ———
   2.25 oz. lost

2. 2.25 oz. lost
   × $16 per oz.
   ———
   $36.00 cost of refining loss

The cost per ounce for refining is about $0.18. The refining fee for the weight in question is:

3. 112.5 oz. fine silver
× $0.18 per oz. refining fee
$20.25 refining fee

Adding the refining loss and fee we get $56.25 as the full cost of refining. The actual value of the silver content is $1800. The dealer's net profit for the purchase and refining of $100 of War Nickels is:

4. $1800.00 full value at spot price
− $56.25 refining loss and fee
$1743.75 value minus refining loss and fee

5. $1743.75
− $ 600.00 dealer's offer
$1143.75 dealer's net profit

The dealer's profit on this transaction is $1,143.75!

28

United States Silver Coins—Dimes, Quarters, Half Dollars—1964 or Older (900 fine)

In the late 1960s the American public did not understand the reasons behind the increased value of their silver coins. They just took advantage of what seemed like a good thing. They could take the change from their pockets or break open their piggy banks. The silver coin they took to the coin merchant's brought them at least 25 percent over face value, which seemed like a good profit. Most people didn't care what happened to the money after they traded it. In fact, many people thought a vast amount of silver coin was being hoarded, awaiting a time when the price would be "right."

These speculators in silver were thought to be watching for an increased numismatic value in the coins.

In reality, the great majority of *all* silver coin minted from 1934 through 1964 has been melted. Clandestine smelting operations were carried out all over the country. As little as $1000 a week to several million dollars in face value were "burned" by a small army of (unassociated) smelters using equipment that ranged from modern to most primitive.

The majority of all U.S. silver coins minted from 1934 through 1964 were melted by these clandestine operators, but the United States government also melted many millions of coins that were withheld from recirculation. The specific price established for a coin issue is based, in part, on the quantity minted; the fewer coins minted, the greater the value of the coin to collectors.

Numismatists who base the value of their collection, at least partly, on the quantity of coins minted, without considering the vast number melted, are using unrealistic figures. Contrary to a widely held notion in the numismatic field, including publishers of coin books and magazines, there are figures available that properly reflect the number of coins melted by the illegal smelters (and the United States government).

To illustrate the way melting of United States silver coins reduced the number of existing coins—presumably affecting the numismatic values—the following example of two specific coin issues is given. This data is taken from the charts found in a planned book, *Manual of Existing United States Silver Coin.*

COIN *(quarter)*	1968 VALUE	QUANTITY MINTED	COINS REMAINING *(after melting)*	1980 VALUE *(EF-40 Cond.)*
1937S	$0.75	1,652,000	1,652,000	$20.00
1958	$0.25	7,235,652	1,809,000	$ 1.50

This graphically illustrates the need for the reevaluation of the numismatic values set on United States silver coins that were subjected to melting in the 1960s. The 1937S and 1958 quarters are equally rare today as a result of the melting craze, but the older coin has an exaggerated value more than thirteen times the value of the just as rare younger coin.

Another little-known fact is that these silver coins, regardless of the specific denomination, be it a dime, quarter, or half dollar, weigh exactly the same at an equal face value (for example, as one dollar).

One half dollar weighs 192.9 gr.
Two half dollars weigh 385.8 gr.

One quarter weighs 96.45 gr.
Four quarters weigh 385.8 gr.

One dime weighs 38.58 gr.
Ten dimes weigh 385.8 gr.

One dollar weighs 385.8 gr.
Ten dollars weigh 3858 gr.

One hundred dollars weigh 38,580 gr.

One thousand dollars weigh 385,800 gr.

These silver coins are 900 fine (90 percent pure silver). One dollar in silver dimes, quarters, or halves weighs a total of .80375 oz. (approximately four-fifths of an ounce). The actual fine weight of the 90 percent silver content of one dollar is .72337 oz. (almost three-fourths of an ounce).

The chart at the end of this chapter shows the total weights of different denominations of silver coins, along with their fine-silver weight.

All the charts following silver-coin categories give weights in ounces, pennyweights, and grams. This is to acquaint you with the different weight systems used on precious-metals scales. You should be familiar with conversions between the different designations, at least insofar as they relate to your dealings. All calculations—in coins, sterling silver, gold, or in whatever you might be interested—must ultimately be converted to ounces to compare the value of your alloy with the spot silver or gold metal price. Because of this, the only figure of direct interest in a deal on silver coins is the *fine-ounce weight* found under the column "Weight of Silver Content."

So you may evaluate the prices quoted by buyers as they most commonly offer them, the following chart takes several set prices and shows the *actual price* the buyer is offering for *one dollar.* If the offer is "15 to 1" some buyers are paying $15 for your $1, while others are paying $15 plus the $1 traded for a total of $16. It appears that most dealers are paying the $15 return, so the chart is based on that figure.

Actual Price per Ounce Being Paid for Silver Coins, One-Dollar Face Value (For United States Dimes, Quarters, Half Dollars—1964 or Older)

15 to 1 ($15) = $20.73 per troy ounce for one dollar
20 to 1 ($20) = $27.64 per troy ounce for one dollar
30 to 1 ($30) = $41.47 per troy ounce for one dollar
40 to 1 ($40) = $55.29 per troy ounce for one dollar

This chart is only a guide; the prices vary so much that it isn't feasible to try to cover all set prices.

To find the rate you are being offered, take the total return you will get on a dollar (if the offer is, say, 20 to 1, check to see if the return is $20 or $21) and divide that price by .72337. The price the buyer is paying is $27.64 per ounce (20 ÷ .72337 = $27.64). If the buyer is paying $27.64 when the silver metal price is, say, $40 per ounce, you can see his markup. Your should *always* check the latest spot silver price in a daily publication so that you can deal intelligently.

The following charts show ounce, pennyweight, and gram comparisons. The pennyweight and gram scales are most commonly used, so to compare the results of these scales with the latest silver metal price, the results must be changed to ounces, as the daily silver metal price is quoted in ounces.

To use these charts you must understand the way to apply them to the daily spot silver price. These charts are as complete as you will need them. You will be dealing in one of the four values: $1, $10, $100 or $1000. When the term "bag" is used, it usually applies to a $1000 bag.

For example, you have coins to sell with a face value of $100. The buyer is offering 20 to 1, and the spot silver price quoted in the daily publication is $40 per ounce. You look at the chart showing $100 and find the weight of the silver content is 72.337 oz. Multiply this by $40, which is the silver metal price: 72.337 oz × $40 =

$2893.48. These silver coins are therefore worth $2893 at a spot price of $40; but the dealer is offering $2000.

The weight of the coins shown on the charts obviously never changes; the spot silver price changes daily, as does the buyer's also. What these charts and systems help you determine is the actual value of the silver in your coin, so you can evaluate the deals offered by the various buyers.

All that is necessary for you to find the actual silver-metal value of your coins is to look at the chart on the line beginning with the face value of your coins. The figure at the far right (under ounces) is the weight of the fine-silver content. Multiply that ounce figure by the daily spot silver price and you have the full value of your coins' silver content. This method is the same for any of the coins covered by charts, all United States and Canadian silver coins.

United States Silver Coin—Dimes, Quarters, Half Dollars—1964 or Older (900 fine)

ACTUAL COIN WEIGHT	WEIGHT OF SILVER CONTENT *(90% of actual coin weight)*
OUNCES	
$1.00 = 0.066 lb. or .80375 oz.	0.0602 lb. or 0.72337 oz.
$10.00 = 0.669 lb. or 8.0375 oz.	0.6028 lb. or 7.2337 oz.
$100.00 = 6.697 lbs. or 80.375 oz.	6.028 lbs. or 72.337 oz.
$1000.00 = 66.97 lbs. or 803.75 oz.	60.28 lbs. or 723.37 oz.
PENNYWEIGHT	
$1.00 = 0.066 lb. or 16.07499 dwt.	0.0602 lb. or 14.46749 dwt.
$10.00 = 0.669 lb. or 160.7499 dwt.	0.6028 lb. or 144.6749 dwt.
$100.00 = 6.697 lbs. or 1607.499 dwt.	6.028 lbs. or 1446.749 dwt.
$1000.00 = 66.97 lbs. or 16,074.9 dwt.	60.28 lbs. or 14,467.4 dwt.
GRAMS (METRIC)	
$1.00 = 0.066 lb. or 24.99984 gr.	0.0602 lb. or 22.49985 gr.
$10.00 = 0.669 lb. or 249.9984 gr.	0.6028 lb. or 224.9985 gr.
$100.00 = 6.697 lbs. or 2499.984 gr.	6.028 lb. or 2249.985 gr.
$1000.00 = 66.97 lbs. or 24,999.8 gr.	60.28 lb. or 22,499.8 gr.

29

United States Silver War Nickels—1942 Through 1945 (350 fine)

It is perhaps in the selling of the 1942 through 1945 War Nickels that the greatest inequities occur. Dealers aware of the actual value of the War Nickel sometimes allow the seller to underrate the value of the silver content of this unusual coin. It is unusual because it is the only nickel made that has no nickel in it. The five-cent coin got its name from the actual metal it contains. Of course, it contains more copper than nickel (75 percent copper, 25 percent nickel), but the name most commonly used for the five-cent piece is nickel. During World War II nickel metal became scarce, and the War Nickel was minted with 56 percent copper, 9 percent manganese, and 35 percent silver.

The face value of the nickel is, of course, half that of the dime: ten dimes make up one dollar, while it takes twenty nickels to equal one dollar. Compare the weight, however. A nickel weighs 77.16 gr.; a dime weighs 38.58 gr.; two dimes weigh 77.16 gr.; ten dimes weigh 385.8 gr.; five nickels weigh 385.8 gr.; and twenty nickels weigh 1543.2 gr.

The dime, however, has a greater silver content (90 percent) than the War Nickel (35 percent). When the actual silver content of ten dimes (0.72337 oz.) is compared with the actual silver content of twenty War Nickels (1.125 oz.), you find that the silver-content value of the War Nickel is 55% greater than the same face value in dimes (1.125 ÷ .72337 = 1.555).

As an example: you want to sell $100 in War Nickels. The buyer is offering 20 to 1 (the same price he is offering for 1964 and older silver U.S. coins). Using the chart, you find $100 worth (face value) of War Nickels has silver content weight of 112.5 oz. The spot silver price quoted in the daily publication is $40 per ounce. Again, the average person might think it was a good deal to be offered the same price for the nickels as for the dimes, quarters, and half dollars. However, if you take the silver-content weight of the nickels (112.5 oz.) and multiply it by $40, which is the spot silver price (112.5

oz. × $40 per oz. = $4,500), you will discover that $100 face value of War Nickels is worth $4,500.

The dealer's offer of 20 to 1 would be figured $100 face value × $20 = $2000. You can see that this profit margin ($2500) is much greater for the buyer than on United States silver coin (in the previous section). Remember, the weights given on the charts never change, only the buyer's offer and the spot silver price change.

United States Silver War Nickel—1942 Through 1945 (350 fine)

ACTUAL COIN WEIGHT		WEIGHT OF SILVER CONTENT *(35% of actual coin weight)*
OUNCES		
$1.00	= 0.267916 lb. or 3.215 oz.	0.09376 lb. or 1.125 oz.
$10.00	= 2.67916 lbs. or 32.15 oz.	0.9376 lb. or 11.25 oz.
$100.00	= 26.7916 lbs. or 321.5 oz.	9.376 lbs. or 112.5 oz.
$1000.00	= 267.916 lbs. or 3215 oz.	93.76 lbs. or 1125 oz.
PENNYWEIGHT		
$1.00	= 0.267916 lb. or 64.3 dwt.	0.09376 lb. or 22.5 dwt.
$10.00	= 2.67916 lbs. or 643 dwt.	0.9376 lb. or 225 dwt.
$100.00	= 26.7916 lbs. or 6430 dwt.	9.376 lbs. or 2250 dwt.
$1000.00	= 267.916 lbs. or 64,300 dwt.	93.76 lbs. or 22,500 dwt.
GRAMS (METRIC)		
$1.00	= 0.267916 lb. or 99.99911 g.	0.09376 lb. or 34.99968 g.
$10.00	= 2.67916 lbs. or 999.9911 g.	0.9376 lb. or 349.9968 g.
$100.00	= 26.7916 lbs. or 9999.911 g.	9.376 lbs. or 3499.968 g.
$1000.00	= 267.916 lbs. or 99,999.11 g.	93.76 lbs. or 34,999.68 g.

30

United States Silver Kennedy Half Dollars—1965 Through 1970 (400 fine)

The Kennedy Half Dollar from 1965 through 1970 is an odd sandwich of assorted metals. The total weight of two coins (one-dollar face value) is .7392 oz. Only 40 percent of that weight is silver, however. At 400 fine, the actual silver-content weight of one-dollar face value is .29568 oz. (.7392 oz. × .40 = .29568 oz.).

This low silver content makes the Kennedy Half Dollar the lowest valued United States silver coin. It was not melted by the illegal smelters of the 1960s because the base value of $3.38 (the point at which the spot silver price would make the Kennedy Half Dollar more valuable for its silver content than as a coin at face value) was higher than the spot silver price of that time. The spot silver price floated around $2.50 per troy ounce, and that was lower than the base price of $3.38.

A peculiarity of the American public also probably contributed to the abundance of the Kennedy half dollars. For some reason, we hoarded half dollars. If a bank or store gave them to us as change, we saved them. This attitude has now changed, and regardless of how we once felt about saving half dollars, many people are now selling them.

A Kennedy half dollar (1965 through 1970) weighs 177.5 gr., 355 grains to one-dollar face value (400 fine). Half dollars, dimes, and quarters from 1964 and older weigh 385.8 gr. for one dollar face value (900 fine).

United States Silver Kennedy Half Dollar—1965 Through 1970 (400 fine)

ACTUAL COIN WEIGHT	WEIGHT OF SILVER CONTENT (*40% of actual coin weight)*
OUNCES	
$1.00 = 0.0616 lb. or .7392 oz.	0.0246 lb. or 0.29568 oz.
$10.00 = 0.616 lb. or 7.392 oz.	0.2464 lb. or 2.9568 oz.
$100.00 = 6.16 lbs. or 73.92 oz.	2.464 lbs. or 29.568 oz.
$1000.00 = 61.6 lbs. or 739.2 oz.	24.64 lbs. or 295.68 oz.
PENNYWEIGHT	
$1.00 = 0.0616 lb. or 14.784 dwt.	0.0246 lb. or 5.9136 dwt.
$10.00 = 0.616 lb. or 147.84 dwt.	0.2464 lb. or 59.136 dwt.
$100.00 = 6.16 lbs. or 1478.4 dwt.	2.464 lbs. or 591.36 dwt.
$1000.00 = 61.6 lbs. or 14,784 dwt.	24.64 lbs. or 5913.6 dwt.
GRAMS (METRIC)	
$1.00 = 0.0616 lb. or 22.99207 g.	0.0246 lb. or 9.19683 g.
$10.00 = 0.616 lb. or 229.9207 g.	0.246 lb. or 91.9683 g.
$100.00 = 6.16 lbs. or 2299.207 g.	2.46 lbs. or 919.683 g.
$1000.00 = 61.6 lbs. or 22,992.07 g.	24.6 lbs. or 9196.83 g.

31

United States Silver Dollars—Through 1935 (900 fine)

For many years, the United States silver dollar was an orphan in the world of trade. If you weren't a collector, you didn't want them. They were too big to use as pocket change, the paper dollar was more convenient. Banks were happy to trade them to you for paper money. The gambling casinos in Nevada were among the few places that used them regularly. "Cartwheels" were a gimmick at the gaming tables.

The silver dollar weighs even more than other United States silver coins. A dollar's worth of dimes, quarters, or half dollars weighs 385.8 gr. A silver dollar weighs 412.5 gr. The silver dollar is 900 fine, the same as other United States silver coins, so at the greater weight, the silver dollar has more value as silver-metal alloy.

Of course, many issues of silver dollars have a high numismatic value, and this value could be greater than the spot silver price. You will have to make that determination by comparing the two values, numismatic and spot silver price. The possibility of a rare coin showing up among your silver dollars is greater than with any other United States silver coin. Pre-1965 dimes, quarters, and half dollars were melted close to extinction in the 1960s. But because many silver dollars had a numismatic value higher than the prevailing spot silver price, very few were "burned" for silver-metal trading.

United States Silver Dollar—Through 1935 (900 fine)

ACTUAL COIN WEIGHT	WEIGHT OF SILVER CONTENT *(90% of actual coin weight)*
OUNCES	
$1.00 = 0.0716 lb. or .859375 oz.	0.0644 lb. or .773437 oz.
$10.00 = 0.7161 lb. or 8.59375 oz.	0.6445 lb. or 7.73437 oz.
$100.00 = 7.161 lbs. or 85.9375 oz.	6.445 lbs. or 77.3437 oz.
$1000.00 = 71.61 lbs. or 859.375 oz.	64.45 lbs. or 773.437 oz.

PENNYWEIGHT		
$1.00	= 0.0716 lb. or 17.1875 dwt.	0.0644 lb. or 15.46875 dwt.
$10.00	= 0.7161 lb. or 171.875 dwt.	0.6445 lb. or 154.6875 dwt.
$100.00	= 7.161 lbs. or 1718.75 dwt.	6.445 lbs. or 1546.875 dwt.
$1000.00	= 71.61 lbs. or 17,187.5 dwt.	64.45 lbs. or 15,468.75 dwt.
GRAMS (METRIC)		
$1.00	= 0.0716 lb. or 26.730 g.	0.0644 lb. or 24.057 g.
$10.00	= 0.7161 lb. or 267.30 g.	0.6445 lb. or 240.57 g.
$100.00	= 7.161 lbs. or 2673.0 g.	6.445 lbs. or 2405.7 g.
$1000.00	= 71.61 lbs. or 26,730 g.	64.45 lbs. or 24,057 g.

32

Canadian Silver Coins—1937 Through 1967 (800 fine)

Canadian silver coins were dealt with by the illegal smelters of the 1960s, but not as heavily as the United States silver coins. The lower silver content (10 percent less) made them less profitable, and United States silver was not too difficult to obtain.

The value of the actual silver content of Canadian silver coins is calculated like United States silver coins. First, check the daily spot silver price. Then look at the chart and find the ounce weight of the amount of money you have to trade (the last column on the right, under "Weight of Silver Content"). Multiply that weight by the spot silver price. The result is the full value of the coin you have.

For example, you have $100 face value worth of Canadian silver coins, and the spot silver price quoted is $35 per ounce. The chart shows that the weight of silver content is 60 oz. Therefore, the full value of the silver content of your coins is $2,100 ($35 per oz. × 60 oz. = $2,100). Compare that with the price the dealer is offering.

To apply the method demonstrated above, the following charts provide the weight of the fine silver in Canadian silver coins. To calculate their value, the only additional information necessary is the daily spot silver price as quoted in a publication.

Canadian Silver Coin—1937 Through 1967 (800 fine)

ACTUAL COIN WEIGHT	WEIGHT OF SILVER CONTENT *(80% of actual coin weight)*
OUNCES	
$1.00 = 0.0625 lb. or 0.75 oz.	0.05 lb. or 0.6 oz.
$10.00 = 0.625 lb. or 7.5 oz.	0.5 lb. or 6 oz.
$100.00 = 6.25 lbs. or 75 oz.	5.0 lbs. or 60 oz.
$1000.00 = 62.5 lbs. or 750 oz.	50 lbs. or 600 oz.
PENNYWEIGHT	
$1.00 = 0.0625 lb. or 15 dwt.	0.05 lb. or 12 dwt.
$10.00 = 0.625 lb. or 150 dwt.	0.5 lb. or 120 dwt.
$100.00 = 6.25 lbs. or 1500 dwt.	5.0 lbs. or 1200 dwt.
$1000.00 = 62.5 lbs. or 15,000 dwt.	50 lbs. or 12,000 dwt.
GRAMS (METRIC)	
$1.00 = 0.0625 lb. or 23.328 g.	0.05 lb. or 18.6624 g.
$10.00 = 0.625 lb. or 233.28 g.	0.5 lb. or 186.624 g.
$100.00 = 6.25 lbs. or 2332.8 g.	5.0 lbs. or 1866.24 g.
$1000.00 = 62.5 lbs. or 23,328 g.	50 lbs. or 18,662.4 g.

33

Base Prices for Silver Coins

Coins, with a specific face value, have a base price. This figure never changes because it is based on the face value of the coin and its weight.

United States silver coin (through 1964) in half dollars, quarters, and dimes weighs 385.8 gr. to one-dollar face value, or, .80375 oz. total weight. At 900 fine (90 percent silver), the actual silver weight is .72337 oz. Divide $1.00 by .72337 and you find the $1.00 has a base price of $1.38 an ounce: ($1.00 ÷ .72337 oz. = $1.38 per oz.). When the spot silver price reached $1.38 per ounce, it was equal to the face value of United States silver coin. Of course, that time is long gone, never to be seen again. When the spot silver price passed the base price value of silver coin, the coin became more valuable as bulk silver metal than as coin.

In the late 1960s, when the majority of United States silver coin was melted down into ingots by clandestine smelters, the operators could buy the coin for 25 percent over face value. With silver metal fluctuating around $2.50 per ounce, even figuring cost of transportation, smelting, and other factors, the profit margin was still over 100 percent. Few businesses could exhibit that kind of financial return. Is it any wonder that most of the United States silver coins were melted in the late 1960s?

Now silver-coin value is tied more closely to the ever-changing spot silver price. It is no longer necessary to melt the coin to make a good profit. Of course, all too often the good profit is made by the buyer; the seller is confused by the figures quoted on silver metal in relation to the prices offered by the buyers. This is the basic purpose of this book. To inform the public.

Base Price for United States and Canadian Silver Coins

COIN	SILVER CONTENT WEIGHT OF $1.00	FINE	BASE PRICE
U.S. War Nickel	1.125 oz.	350 fine	$0.89
U.S. Silver Dollar (thru 1935)	0.77343 oz.	900 fine	$1.29
U.S. Silver Coin (thru 1964)	0.72337 oz.	900 fine	$1.38
Canadian Silver Coin (1937 thru 1967)	0.6 oz.	800 fine	$1.67
U.S. Kennedy Half (1965 thru 1970)	0.29568 oz.	400 fine	$3.38

* This chart is arranged in the order of the value of the coin—the lower the base price, the more valuable the coin.

The method of finding a base price is to divide $1.00 by the actual silver content weight of $1.00. For example, the base price of the U.S. silver coin (through 1964) has a silver content weight of .72337 of an ounce: $.72337\overline{)1.0000}$ = 1.3824 = $1.38 base price

34

Note on The United States Copper Penny

Although the United States penny is not a precious metal, the base price at which it becomes profitable to sell the coin as bulk metal is appropriate to mention in this book. Because the penny is a non-precious metal, transactions concerning it are based on the avoirdupois system, not on the troy.

The United States copper penny weighs 48 gr. and is 95 percent copper, that is, 45.6 gr. copper and 2.4 gr. tin/zinc. One hundred pennies ($1.00) have a net weight of 45.6 gr., or 0.65142 lb. of pure copper. Divide $1.00 by 0.65142 = $1.54. The base price of copper United States pennies is $1.54 per avoirdupois pound. When the quoted price rises above $1.54 per pound, the United States copper penny becomes a candidate for bulk trading.

Equipment and Supply Sources

For persons interested in a precious-metals business, the following firms can furnish acid test kits and other items related to the operation:

Gamzon Bros. Inc.
21 West 46th St.
New York, N.Y. 10036

I. Shor Company
50 West 23rd St.
New York, N.Y. 10010

Friedham Tool Supply Co.
412 West 6th St.
Los Angeles, Ca. 90014

If you are interested in purchasing scales made by either of the two firms mentioned in this book, I suggest you write the manufacturers to obtain the name of the nearest firm selling their scales. Their addresses are:

Brinkmann Instruments, Inc.
Cantiaque Road
Westbury, N.Y. 11590

Ohaus Scale Corp.
29 Hanover Rd.
Florham Park, N.J. 07932

Afterword

RADIOACTIVE GOLD JEWELRY

A problem not normally associated with gold, or any precious metal, is the possibility that wearing a piece of jewelry could be dangerous to your health. Newspaper headlines in the early eighties brought the problem out in the open, but it is a matter of record that radioactive gold jewelry was detected as early as 1935.

Contaminated gold, mostly in the form of rings, has been uncovered in one main geographic area—western New York and Pennsylvania.

If you own a ring, bracelet, or any piece of jewelry that has ever caused you skin irritation or other medical problems, take the jewelry to a center with facilities for testing for radioactivity.

The source of the radioactive gold is believed by experts to have been radon seeds, very small gold capsules used mainly to treat specific forms of cancer. Although such medical use has been largely discontinued, there are still at least three operating radon plants left from a one-time total of over thirty. Modern control of spent radon seeds is tight, but with the current high price of gold, diligence must be maintained. When gold was $35.00 an ounce or less, each radon seed was worth approximately 3 cents; but it is now more profitable for persons with warped values to attempt to pass off spent radon seeds to gatherers of precious metals.

Fortunately for the public, most major refineries routinely check their gold for radioactivity. Many smaller smelters, however, and many individual jewelers and manufacturers refine their own gold, and it is this potential weak point in screening which must be strengthened. The federal government, through the United States Nuclear Regulatory Agency, has set and maintained stringent controls over the manufacturing, handling, and use of radioactive precious metals. Many states have taken firm and enlightening stands in meeting this problem, not the least being New York State through its Department of Health. Today, most of the known contaminated rings in New York are safely under the control of its Department of Health. The conscientious efforts of public officials everywhere should result in isolation of the problem, then, over a necessary period of time, its elimination. Because this health problem to date has been confined to the western New York-Pennsylvania area, it can be assumed that the threat to public health is under control. There is one disturbing factor, however: If jewelers in the western New York-Pennsylvania area used contaminated gold in the thirties and forties, it is possible that the widespread medical use of radon seeds at that time led to the as yet undetected use of contaminated gold in jewelry in other areas. Luckily, with proper equipment, contamination is easily detected and can be brought under control.

Index